FANTASTIC NUMBERS

AND WHERE

TO FIND THEM

FANTASTIC NUMBERS

AND WHERE

TO FIND THEM

A Cosmic Quest from Zero to Infinity

ANTONIO PADILLA

Farrar, Straus and Giroux | New York

Farrar, Straus and Giroux
120 Broadway, New York 10271

Copyright © 2022 by Antonio Padilla
All rights reserved
Printed in the United States of America
Originally published in 2022 by Allen Lane, Great Britain
Published in the United States by Farrar, Straus and Giroux
First American edition, 2022

Library of Congress Cataloging-in-Publication Data
Names: Padilla, Antonio, 1975– author.
Title: Fantastic numbers and where to find them : a cosmic quest from zero to
    infinity / Antonio Padilla.
Description: First edition. | New York : Farrar, Straus and Giroux, 2022. |
    "Originally published in 2022 by Allen Lane, Great Britain" | Includes
    bibliographical references and index.
Identifiers: LCCN 2022009710 | ISBN 9780374600563 (hardcover)
Subjects: LCSH: Mathematical physics—Popular works. | Number theory—
    Popular works.
Classification: LCC QC20 .P2675 2022 | DDC 530.15—dc23/eng20220517
LC record available at https://lccn.loc.gov/2022009710

Our books may be purchased in bulk for promotional, educational,
or business use. Please contact your local bookseller or the Macmillan
Corporate and Premium Sales Department at 1-800-221-7945,
extension 5442, or by email at MacmillanSpecialMarkets@macmillan.com.

www.fsgbooks.com
www.twitter.com/fsgbooks • www.facebook.com/fsgbooks

10 9 8 7 6 5 4 3 2 1

To my girls
(who called me Gilderoy)

Contents

Little Numbers

Infinity

FANTASTIC NUMBERS

AND WHERE

TO FIND THEM

A Chapter That's Not a Number

The number lay there, brazen, taunting me from the tatty piece of paper that sat neatly on the ancient oak table: zero. I'd never scored zero in a maths test before but there was no mistaking my mark. The number was scrawled aggressively in red at the top of the coursework I'd handed in a week or so earlier. This was in my first term as a mathematics undergraduate at Cambridge University. I imagined the ghosts of the university's great mathematicians whispering their contempt. I was an imposter. I didn't know it at the time, but that coursework would prove to be a turning point. It would change my relationship with both maths and physics.

The coursework had involved a mathematical proof. These usually begin with some assumptions and, from there, you infer a logical conclusion. For example, if you assume that Donald Trump was both orange and President of the United States, you may infer that there has been an orange President of the United States. My coursework had nothing to do with orange presidents, of course, but it did involve a series of mathematical statements that I'd connected with a clear and consistent argument. The Cambridge don agreed – all the arguments were there – but he had still given me a zero. It turned out his issue was with how I'd laid it all out on the tatty piece of paper.

I was frustrated. I'd done the hard part in figuring out the solution to the coursework problem, and his complaint seemed petty. It was as if I'd scored a spectacular goal, only for the don to check with the Video Assistant Referee and rule it out for a marginal offside. But I now know why he did it. He was trying to teach me about rigour, trying to instil the mathematical pedantry that is an essential part of a mathematician's toolkit. Reluctantly, I became a pedant, but I also realized then that I needed a little more from mathematics. I needed it to have personality. I'd always loved numbers, but I wanted to bring them to life – to give them a purpose – and for that I found that I needed physics. That is what this book is all about – the personality of numbers shining through in the physical world.

Take Graham's number as an example. This is a leviathan, a number so large that it once had pride of place in the *Guinness Book of World Records* as the largest number ever to appear in a mathematical proof. It is named after the American mathematician (and juggler) Ron Graham, who was wonderfully pedantic in making mathematical use of it. But his pedantry is not what brings Graham's number to life. What brings it to life – or perhaps more accurately, death – is physics. You see, if you were to try and picture Graham's number in your head – its decimal representation written out in full – your head would collapse into a black hole. It's a condition known as *black hole head death* and there is no known cure.

In this book, I'm going to tell you why.

In fact, I'm going to tell you more than why. I'm going to take you to a place where you will question things you'd always assumed to be true. This journey through *Fantastic Numbers* will begin with the biggest numbers in the universe and a quest to understand what is known as *the holographic truth*. Are three dimensions just an illusion? Are we trapped inside a hologram?

To understand this question, punch the air around you. You should probably make sure you aren't sitting too close to anyone, but punch forwards and backwards, left and right, and up and down. You can punch your way through three dimensions of space, three perpendicular directions. Or can you? The holographic truth asserts that one of these dimensions is a fake. It is as if the world is a 3D movie. The real images are trapped on a two-dimensional screen, but when the audience puts on their glasses a 3D world suddenly emerges. In physics, as I will explain in the first half of this book, the 3D glasses are provided by gravity. It is gravity that creates the illusion of a third dimension.

It was only by taking gravity to its extreme that we became aware of its sorcery. But then this is a book of extremes. Our quest to understand the holographic truth begins, inevitably, with Albert Einstein, his genius, the perverse brilliance of relativity and the underlying structure of space and time. Of course, I have a number for his genius: 1.00000000000000858. And yes, I'm calling this a big number. I imagine you are sceptical, but hopefully I'll convince you that it is a huge number, at least if you think about the physics it represents: one man's ability to meddle with time. To really understand why, we'll need to run alongside the legendary Jamaican sprinter Usain Bolt. We'll need to plunge to the depths of the Pacific Ocean, to the deepest part of the Mariana Trench. We'll have to go to the edge of physics, dancing dangerously close to a monstrous black hole as it guzzles greedily on the stars and planets at the centre of a distant galaxy.

But relativity and black holes are just the beginning. To find the holographic truth, we will need four more leviathans – genuine numerical gargantua that come to life whenever they collide with the physical world. From a googol to a googolplex, from Graham's number to TREE(3), these are the titanic numbers that will appear to break physics. But the truth is they will guide us in our understanding. They will teach us the meaning of entropy, so often misunderstood, which describes the turbulent physics of secret and disorder. They will introduce us to quantum mechanics, the lord of the microworld, where nothing is certain and everything is a game of chance. The story will be told with tales of doppelgängers in far-off realms and warnings of a cosmic reset, when everything in our universe returns, inevitably, to the way it once was.

In the end, in this land of giants, we will find it: a holographic reality. Our reality.

I am a child of the holographic truth. It was an idea that took off around the time I scored zero in my coursework, although I knew nothing about it back then. By the time I started my doctorate about five years later, it was fast becoming the most important idea to be developed in fundamental physics in almost half a century. Everyone in physics seemed to be talking about it. Everyone is *still* talking about it. They are asking deep and important questions about black holes and quantum gravity and, in the holographic truth, they are finding answers.

There was something else everyone was talking about back then, as we were getting ready to usher in a new millennium: the mystery of our finely tuned and unexpected universe. You see, ours is a universe that simply should not exist. It's a universe that has let us live, that has given us a chance of survival, against all the odds. It's where we will go in the second part of this book, guided not by leviathans but by the mischief-makers – the little numbers.

Little numbers betray the unexpected. To understand this, imagine me winning *The X Factor*. I cannot stress how unexpected this would be, because I'm a terrible singer, so awful that in a high-school musical I was asked by the teachers to stand away from the microphones. With this in mind, I would say that the probability of me winning a national singing competition is somewhere in the region of the following number:

$$\frac{1}{number\ of\ people\ living\ in\ UK} \approx 0.000000015$$

That's quite a small number. Then again, my success would be quite unexpected.

Our universe is even more unexpected. With little numbers as our guide, we will explore this unexpected world. They don't get any smaller than zero, the ugly number that spread its scorn all over my university coursework. The contempt I felt for zero on that particular day has been repeated throughout history. Of all the numbers, zero has been the most unexpected and the most feared. This is because it was identified with the void, with the absence of God and with evil itself.

But zero is neither evil nor ugly; in fact, it is the most beautiful number there is. To understand its beauty we must understand the elegance of the physical world. To a physicist, the most important aspect of zero is its symmetry under a change of sign: minus zero is exactly the same as plus zero. It is the only number with this property. In nature, symmetry is the key to understanding why things vanish, why they equate to the mythical zero.

Things start to get confusing when we encounter small yet non-zero numbers, since they reflect the absurdity of the way the universe seems to be set up as well as our struggles in trying to make sense of it. We will tell this particular story through two disturbingly small numbers, one that betrays the mysteries of the microworld and the other the mysteries of the cosmos. Through the prism of the alarmingly little 0.000000000000001, we enter the subatomic world of particle physics: gluons, muons, electrons and taus, dancing around in random abandon. And there we will find the Higgs boson – the so-called God particle – tying them all together. The Higgs boson was discovered in a whirl of particle excitement in the summer of 2012. It was heralded as a triumph for theory and experiment, ending a near-fifty-year wait for confirmation of the particle's existance. But in among the fanfare was a secret: something didn't quite add up. It turns out that the Higgs boson is far too light, 0.0000000000000001 times lighter than it should be. That's a very little number. It tells us that the microworld lurking within you and around you is very unexpected indeed.

When we get to the number 10^{-120}, we will see that the cosmos is even more unexpected. We see it in the light of distant stars exploding out of existence. The light is dimmer than expected, suggesting that the stars are further away than we'd originally thought. It points to an unexpected universe whose expansion is speeding up, the space between galaxies growing at an accelerated rate.

Most physicists suspect that the universe is being pushed by the vacuum of space itself. That might sound strange – how could empty space push galaxies apart? The truth is that empty space is not so empty, not when you factor in quantum mechanics. It is filled with a bubbling broth of quantum particles frantically popping in and out of existence. It is this broth that pushes on the universe. We can even calculate how hard it pushes, and that's when things start to fall apart. As we will see, the universe is pushed only by a tiny amount, a fraction of what we expect based on our current understanding of fundamental physics. The fraction is just 10^{-120}, less than one part in a googol. This tiny number is the most spectacular measure of our unexpected universe.

It turns out that we are incredibly fortunate. If the universe had been pushed as hard as our calculations suggest it should have been, it would have pushed itself into oblivion, and the galaxies, stars and planets would never have formed. You and I would not exist. Our unexpected universe is a blessing but also a cosmic embarrassment, given our inability to properly understand it. It's a puzzle that has dominated my entire career and continues to dominate it.

But there is something beyond all of this, something deeper and even more profound than our quest for a holographic truth or to understand our unexpected universe. To discover it, we will need our final number, a number that isn't always a number and, at the same time, is many different numbers. It is the number that has confounded mathematicians throughout history, driving some to ridicule and others to madness: infinity.

As the German mathematician David Hilbert, a father to both quantum mechanics and relativity, once said: 'The infinite! No other question has ever moved so profoundly the spirit of man.' Infinity will be our gateway to the Theory of Everything – the theory that underpins all of physics and could one day describe the creation of the universe.

It was Georg Cantor, an outcast of German academia in the late nineteenth century, who dared to climb the infinite tower, layer upon layer, to infinities beyond the infinite. As we will see, he developed the careful language of sets, collections of this and that, that enabled him to rigorously reach into the heavens, to categorize one layer of infinity after another. Of course, he was driven quite mad, wrestling with numbers that seem to have more in common with the divine than with the physical realm. But what of the physical realm? Does it contain the infinite? Is the universe infinite?

The quest to understand physics at its most fundamental, at its most microscopically pure, is the quest to conquer its most violent infinities. These are the infinities we encounter at the core of a black hole, at the so-called singularity, where space and time are infinitely torn and twisted and gravitational tides are infinitely strong. These are also the infinities we encounter at the moment of creation, at the instant of the Big Bang. The truth is these infinities are yet to be conquered and fully understood, but there is promise in a cosmic symphony – a Theory of Everything where particles are replaced with the tiniest strings, vibrating in perfect harmony. As we will discover, the song of the strings doesn't just echo through space and time, it *is* space and time.

The big, the small and the frightfully infinite. Together these are the *Fantastic Numbers*, numbers with pride and personality, numbers that have taken us to the edge of physics, revealing a remarkable reality: a holographic truth, an unexpected universe, aTheory of Everything.

I think it's time to find those numbers.

Big Numbers

1.000000000000000858

A BOLT OF RELATIVITY

Among all the usual football-related paraphernalia there was something different under the Christmas tree that year. It was a *dictionary*, one of those classic Collins ones that could serve as a barricade should the need ever arise. I'm not sure why my mum and dad thought fit to buy their ten-year-old son a dictionary when, at that stage, I had shown relatively little interest in words. In those days, I had two passions in life: Liverpool Football Club and maths. If my parents thought this present would broaden my horizons, they were sorely mistaken. I considered my new toy and decided I could at least use it to look up massive numbers. First I searched for a billion, then a trillion, and it wasn't long before I discovered a 'quadrillion'. This game went on until I happened upon the truly magnificent 'centillion'. Six hundred zeroes! That was in old English, of course, before we embraced the short-scale number system. Nowadays a centillion has a less inspiring 303 zeroes, just as a billion has nine rather than twelve.

But this was as far as it went. My dictionary didn't contain a googolplex or Graham's Number or even TREE(3). I would have loved them back then, these leviathans. Fantastic numbers like these can take you to the brink of our understanding, to the edge of physics, and reveal fundamental truths about the nature of our reality. But our journey begins with another big number, one that was also absent from my Collins dictionary: 1.000000000000000858.

I imagine you're disappointed. I've promised you a ride with numerical leviathans, but this number doesn't seem to be very big at all. Even the Pirahã people of the Amazon rainforest can name something bigger, and their number system includes only *hoí* (one), *hói* (two) and *báagiso* (many). To make matters worse, it's not even a very *pretty* or elegant number like pi or root 2. In every conceivable sense, this number appears to be remarkably unremarkable.

This is all true until we start to think about the nature of space and time and the extremes of our human interactions with them. I chose this particular number because it's a *world record* for its size, revealing the limit of our physical ability to meddle with the properties of time. On 16 August 2009 Jamaican sprinter Usain Bolt managed to slow his clock by a factor of 1.0000000000000000858. No human has ever slowed time to such an extent, at least not without mechanical assistance. You may remember this event differently, as the moment when the 100-metre world record was shattered at the athletics world championships in Berlin. Watching in the stadium that day were Wellesley and Jennifer Bolt, whose son hit a top speed of 27.8mph (12.42m/s) between the 60- and 80-metre mark of the race. For each second experienced by their son in those moments, Wellesley and Jennifer would experience a little more: 1.0000000000000000858 seconds, to be precise.

To understand how Bolt was able to slow time, we need to accelerate him up to the speed of light. We need to ask what would happen if he were able to catch up with it. You can call this a 'thought experiment' if you like, but don't forget that Bolt managed to break three world records at the Beijing Olympics, fuelled by a diet of chicken nuggets. Imagine what he could have achieved if he ate properly.

To have any hope of catching light, we must assume that it travels at a finite speed. That is already far from obvious. When I told my daughter that the light from her book did not reach her eye in an instant she was immediately very sceptical and insisted on conducting an experiment to find out if it was really true. I typically get a nosebleed whenever I stray too close to experimental physics, but my daughter seems to have acquired more of a practical skill set. She set things up as follows: turn the bedroom light off, then turn it on again and count how long it takes for the light to reach you. This is exactly the same sort of experiment carried out by Galileo and his assistant using covered lanterns four hundred years ago. Like my daughter, he concluded that the speed of light 'if not instantaneous ... is extraordinarily rapid'. Rapid, but finite.

By the mid-nineteenth century physicists such as the wonderfully named Frenchman Hippolyte Fizeau were beginning to home in on a reasonably accurate – and finite – value for the speed of light. However, to properly understand what it would mean to catch up with light, we need to first focus on the remarkable work of the Scottish physicist James Clerk Maxwell. It will also illustrate the beautiful synergy that exists between maths and physics.

By the time Maxwell was considering the behaviour of electricity and magnetism there were already hints that they could be two different sides of the same coin. For example, Michael Faraday, one of England's most influential scientists, despite his lack of formal education, had previously discovered the law of induction, showing that a changing magnetic field produced an electric current. The French physicist André-Marie Ampère had also established a connection between the two phenomena. Maxwell took these ideas and the corresponding equations and tried to make them mathematically rigorous. But he noticed an inconsistency – Ampère's law, in particular, defied the rules of calculus whenever there was a flux of electric current. Maxwell drew analogies with the equations that governed the flow of water and proposed an improvement on what Ampère and Faraday had to offer. Through mathematical reason, he found the missing pieces of the electromagnetic jigsaw and a picture emerged of unprecedented elegance and beauty. It is this strategy, pioneered by Maxwell, that pushes the frontiers of physics in the twenty-first century.

Having established his mathematically consistent theory, unifying electricity and magnetism, Maxwell noticed something magical. His new equations admitted a wave solution, an *electromagnetic wave*, where the electric field rises and falls in one direction and the magnetic field rises and falls in the other. To understand what Maxwell found, imagine two sea snakes coming straight for you on a scuba dive. They are travelling along a single line in the water, the 'electric' snake slithering up and down, the 'magnetic' snake slithering left and right, and to make matters worse, they are charging towards you at 310,740,000m/s. The last bit of the analogy might be the most terrifying, but it is also the most remarkable part of Maxwell's discovery. You see, 310,740,000m/s really was the speed that Maxwell calculated for his electromagnetic wave – it just popped out of his equations like a mathematical jack-in-the-box. Curiously enough, that figure was also very close to the estimates for the *speed of light* that had been measured by Fizeau and others. Remember: as far as anyone was aware at the time, electricity and magnetism had nothing to do with light, and here they were, apparently consisting of waves travelling at the same speed. Modern measurements of the speed of light through a vacuum place its value at 299,792,458m/s, but the parameters of Maxwell's equations are also known to a greater accuracy and the miraculous coincidence survives. Because of this coincidence, Maxwell realized that light and electromagnetism *had* to be one and the same thing: an astonishing connection between two apparently separate properties of the physical world revealed by mathematical reason.

It gets better. Maxwell's waves didn't just include light. Depending on their frequency of oscillation or, in other words, the rate at which the sea snakes slither from side to side, the wave solutions described radio waves, X-rays and gamma rays, and although the frequencies were different, the speed at which they moved was always the same. It was the German physicist Heinrich Hertz who actually measured radio waves, in 1887. When he was quizzed about the implications of his discovery, Hertz humbly replied, 'It is of no use whatsoever. This is just an experiment that proves Maestro Maxwell was right.' Of course, whenever we tune a radio station to the desired frequency, we are reminded of the real impact of Hertz's discovery. But even if he underplayed his own importance, Hertz was right to describe Maxwell as a maestro. He was, after all, conductor of the most elegant mathematical symphony in the history of physics.

Before Albert Einstein revolutionized our understanding of space and time, it had been widely assumed that waves of light require a medium through which to propagate, much in the way that waves on the ocean need to propagate through a body of water. The imagined medium for light was known as the *luminiferous aether*. Let's assume, for a moment, that the aether is real. If Usain Bolt were to catch up with light, he would have to travel through the aether at 299,792,458m/s. If he *did* get up to speed, then once he is running alongside the light ray, *what would he actually see*? The light would no longer be moving away from him so it would just appear as an electromagnetic wave oscillating up and down and left and right but not actually going anywhere. (Imagine the sea snakes slithering to and fro but ultimately staying in the same place in the ocean.) But there is no obvious way to adapt Maxwell's laws to allow for this sort of wave, which suggests that the laws of physics would have to be radically different for the supercharged version of the Jamaican sprinter.

This is unsettling. When Einstein drew the same conclusions, he knew that something had to be wrong with this idea of catching up with light. Maxwell's theory was much too elegant to abandon just because somebody happened to be moving quickly. Einstein also needed to find a way of taking into account the strange results of an experiment carried out in Cleveland, Ohio, in the spring of 1887. Two Americans, Albert Michelson and Edward Morley, had been trying to find the speed of the Earth through the aether using some clever arrangement of mirrors, but the answer kept coming out as zero. If correct, this would have meant that the Earth, unlike almost all of the other planets in the solar system and beyond, just so happened to be running right alongside this space-filling aether, at *exactly* the same speed and in *exactly* the same direction. As we will come to appreciate later in this book, coincidences like that don't tend to happen without good reason. The simple truth is that there is no aether – and that Maestro Maxwell is *always* right.

Einstein proposed that Maxwell's laws, or indeed any other physical laws, would *never* change, no matter how quickly you move. If you were locked away in a windowless cabin on a ship, there would be no experiment you could do to detect your absolute velocity because *there is no such thing as absolute velocity*. Acceleration is a different story, and we'll come to that, but as long as the captain of the ship set sail at constant velocity relative to the sea, be it at 10 knots, 20 knots or close to the speed of light, you and your fellow experimenters in the cabin would be blissfully unaware. As for Usain Bolt, we now know that his chase would be futile. He would never catch the light ray because Maxwell's laws can never change. No matter how fast he ran, he would always see the light as if it were moving away from him at 299,792,458m/s.

This is all very counterintuitive. If a cheetah runs across the plain at 70mph and Bolt chases after it at 30mph, then everyday logic would suggest that the cheetah will extend its lead on Bolt by 40 miles every hour, simply because its relative speed is calculated as 70mph – 30mph = 40mph. But when we are talking about a ray of light travelling at 299,792,458m/s across the plain, it doesn't matter how fast Bolt runs, the ray of light will still move relative to Bolt at 299,792,458m/s. Light will always travel at 299,792,458 m/s,[1] relative to the African plain, relative to Usain Bolt, relative to a herd of panicking impala. It really doesn't matter. We can sum it up in a single tweet:

The speed of light is the speed of light.

Einstein would have liked this. He always said that his ideas should have been described as 'the Theory of Invariance', focusing on their most important features: the invariance of the speed of light and the invariance of the laws of physics. It was another German physicist, Alfred Bucherer, who coined the phrase 'the Theory of Relativity', ironically while criticizing Einstein's work. We call it the *special* theory of relativity in order to emphasize the fact that all of the above applies only to motion that is uniform, in other words, with no acceleration. For accelerated motion, like a Formula One driver hitting the gas or a rocket being fired into space, we need something more general and more profound – Einstein's *general* theory of relativity. We'll get to that in detail in the next section, when we plunge to the bottom of the Mariana Trench.

For now, let's stick with Einstein's special theory. In our example, Bolt, the cheetah, the impala and the ray of light are all assumed to be moving with constant velocity relative to one another. Those velocities may differ, but they don't change with time, and the most important thing is that, despite those differences, everyone sees the light ray speeding away at 299,792,458m/s. As we have already seen, this universal perception of the speed of light certainly contradicts our everyday understanding of relative velocities, in which one velocity is subtracted from another. But this is only because you aren't exactly used to travelling around at speeds close to the speed of light. If you were, you would look at relative velocities very differently.

The problem is time.

You see, all along you have been assuming that there is a big clock in the sky that tells us all what time it is. You might not *think* you are assuming this, but you are, especially when you start subtracting relative velocities using what you believe to be common sense. I'm sorry to disappoint you, but this absolute clock is a fantasy. It doesn't exist. All that ever matters is the clock on your wristwatch, or on my wristwatch, or the clock ticking along on a Boeing 747 as it flies across the Atlantic. Each and every one of us has our own clock, our own time, and these clocks don't necessarily agree, especially if someone is hurtling around close to the speed of light.

Let's suppose I jump aboard a Boeing 747. Taking off from Manchester, by the time it reaches the British coast at Liverpool, the aircraft is cruising along at several hundred miles per hour. I decide to bounce a ball a couple of metres across the floor of the cabin, to the slight irritation of the other passengers. My sister, Susie (who happens to live in Liverpool), is on the beach as the plane flies over and, from her perspective, the ball moves considerably further, some two hundred metres or more. At first glance, this doesn't seem to require any major revision of our everyday concept of time. After all, the ball just gets a piggyback from the fast-moving aircraft – of course she sees it move further. But now let's play a similar game with light. I switch on a light on the floor of the cabin, shining a ray vertically upwards, perpendicular to the direction of travel of the aeroplane. In a very short time, I see the light climb up to the cabin ceiling. If Susie were able to see inside, she would see the light travel along a diagonal, rising from floor to ceiling but also moving horizontally with the aircraft.

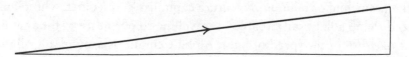

Trajectory of light ray as seen by Susie on the beach.

Her diagonal distance is longer than the vertical distance I measured. That means that she saw the light travel further than I did and yet she saw it travelling at the *same speed*. That can mean only one thing: for Susie, the light took longer to complete its journey; from her perspective, the world inside the aircraft must be ticking along in slow motion. This effect is known as *time dilation*.

The amount by which time is slowed depends on the relative speed, of me with respect to my sister, of Usain Bolt with respect to his parents in Berlin. The closer you are to the speed of light, the more you slow down time. When Bolt was running in Berlin, he hit a top speed of 12.42m/s, and time was slowed by a factor of 1.0000000000000008858.[2] That's the record for human relativity.

There is another consequence of slowing down time – you age more slowly. For Usain Bolt, it turns out he aged about 10 femtoseconds less than everyone else in the stadium during the race in Berlin. A femtosecond doesn't seem like much – it's only a millionth of a billionth of a second – but still, *he aged less*, so when he came to rest he had leapt into the future, albeit very slightly. If you aren't much of a runner, you can take advantage of some mechanical assistance to slow down time and, chances are, you will do even better. Russian cosmonaut Gennady Padalka spent 878 days, 11 hours and 31 minutes in space aboard both the Mir Space Station and the International Space Station, orbiting the Earth at speeds of around 17,500mph. Over the course of these missions, he managed to leap forward a record 22 milliseconds in time compared to his family at home on Earth.*

But you don't have to be a cosmonaut to time-travel in this way. A cabbie driving through the city for forty hours a week for forty years will be a few tenths of a microsecond younger than he would have been had he just stayed put. If you aren't impressed by microseconds and milliseconds, consider what could happen to any bacteria hitching a ride aboard the *Starshot* mission to Alpha Centauri. *Starshot* is the brainchild of billionaire venture capitalist Yuri Milner, who plans to develop a light sail capable of travelling to our nearest star system at *one fifth* of the speed of light. Alpha Centauri is around 4.37 light years away, so we would have to wait more than twenty years on Earth for it to complete its journey. For the light sail and its bacterial stowaway, however, time would slow down to such an extent that the journey would take less than nine years.

At this point, you may have spotted something suspicious. Travelling at one fifth of the speed of light for nine years, the intrepid bacterium

---

* This number also takes into account the negative effect due to his high altitude and weak gravity, effects that will be discussed later on in this chapter.

will cover less than two light years – which is less than half the distance to Alpha Centauri. It's the same with Usain Bolt. I told you that he ran for 10 femtoseconds less than you might have thought, which suggests he didn't actually run as far. And it's true – he didn't. From Bolt's perspective, the track was moving relative to him at 12.42m/s and so it must have shrunk by around 86 femtometres, which is the width of around fifty protons. You could even argue that he didn't quite finish the race. For the bacterium, the space between Earth and Alpha Centauri was moving very quickly and as a result it shrank to less than half its original length. This shrinking of space, or of the racetrack in Berlin, is known as *length contraction*. So you see, running will not only make you age less, it can also help you look thinner. If you ran close to the speed of light, anyone watching would see you flatten out like a pancake, thanks to the shrinking of the space you occupy.

There is something else you should be worried about. I just said that the track was moving relative to Usain Bolt at 12.42m/s. That means that his parents were *also* moving, relative to their son, at exactly the same speed. But given everything we have established so far, this means that Bolt would have seen his parents' clocks slow down, which is very weird, because I already told you that *they* also saw *his* clock slow down. In fact, this is exactly what happens: Wellesley and Jennifer see their son in slow motion (!), and Bolt sees *them* in slow motion. But here's the *really* troubling part: I also said that Bolt managed to finish the race 10 femtoseconds younger than he would have been had he stood still. Couldn't we flip things around and look at it from Bolt's perspective? Time is ticking more slowly for his parents, so couldn't it be they who age less? It seems we have a paradox. This is known as the *twin paradox*, because of the narrative usually used to explain it, but unfortunately Usain Bolt doesn't have a twin. No matter. The truth is that it is Bolt who ages less, who stays that little bit younger. But why him and not his parents?

In order to answer this question, we have to consider the role of acceleration. Remember, everything we have discussed so far applies to uniform motion when there is *no* acceleration. In those moments where Bolt is running at a constant 12.42m/s, he and his parents are what we would call *inertial*. This is just some fancy jargon that says they aren't accelerating – they don't feel any additional force speeding them up or slowing them down. Whenever this is the case, the laws of special relativity apply and so Bolt will see his parents in slow motion, and vice versa. However, Bolt doesn't run at a constant speed for the entire race: he accelerates from zero up to his top speed before slowing down again at the end. In those periods when he is accelerating or decelerating he is *not inertial*, in contrast to his parents. Accelerated motion is a very different beast. For example, locked away in a cabin of a ship, you would certainly be able to tell if the ship was accelerating because you would *feel* the force acting on your body. Too large an acceleration could even kill you. Bolt was never at risk of death, but his acceleration and deceleration were enough to break the equivalence between him and his parents. This asymmetry takes care of the paradox – a more detailed analysis, carefully factoring in Bolt's accelerated motion, reveals that of all the protagonists it was indeed Bolt who aged that little bit less.

It is important to realize that this isn't just some fun with equations. These are real effects that have been *measured*. Fast-moving atomic clocks have been seen to tick more slowly than their stationary counterparts, 'ageing less', just as Usain Bolt did in Berlin. Further evidence comes from a microscopic particle called the muon and its apparent stay of execution. The muon is very much like the electrons you find orbiting the nucleus of an atom, but it's about two hundred times heavier and it doesn't live anywhere near as long. After about two millionths of a second it decays into an electron and some little neutral particles called neutrinos. There is an experiment at Brookhaven National Laboratory in New York in which muons are accelerated around a 44-metre ring at 99.94 per cent of the speed of light. Given their short life span, you would expect the muons to complete only 15 laps; somehow, though, they make it around 438 times. It's not that they live any longer – if you were travelling alongside one at the same speed, you would still see it decay after two millionths of a second – but then you would also see the circumference of the ring shrink to $\frac{1}{29}$ of its original size. The muon gets around 438 times because it has less distance to travel, thanks to length contraction.

Length contraction and time dilation help us understand why nothing – not even Usain Bolt – can travel faster than light. As he gets closer and closer to light speed, Bolt's time appears to slow to a standstill and the distances he encounters shrink to nothing. How can time slow down any more? How can distances shrink to any less? There is simply nowhere to go. The speed of light now presents itself as a barrier and the only reasonable conclusion is that no one can go any faster.

As he accelerates towards the speed of light, Bolt takes on more and more calories to try and accelerate faster and faster. The speed of light looms large as a barrier not to be crossed and so eventually his speed begins to plateau and his acceleration slows down. The closer he gets to the speed of light, the harder it becomes. His resistance to acceleration or, in other words, his *inertia*, just gets larger and larger. That is the problem with trying to accelerate up to the speed of light: inertia blows up to infinity.

But *where* is this inertia coming from? Well, the only thing that Bolt is bringing into the system is energy, and so that energy must be the source of Bolt's extra inertia. Energy never goes away, it just changes how it looks, moving from one form to another. So, inertia must be a form of energy, and this must still be true *even when Bolt is resting*. The cool thing is that for a resting Bolt, we know exactly what his inertia is: it's just his mass, because the heavier he is, the harder he is to move. Mass and energy become one and the same or, as Einstein put it[3]: $E = mc^2$. The terrifying thing about this formula is quite how much energy (E) you can get from mass (m), thanks to the enormous value for the speed of light (c). A resting Usain Bolt weighs around 95 kilograms, and if you were to convert all of that mass into energy it would be the equivalent of 2 billion tons of TNT. That is more than a hundred thousand times the energy released by the Hiroshima bomb.

Now let's talk about spacetime.

Wait. What? Where did that come from? The truth is we've been talking about spacetime all along. Length contraction. Time dilation. In the vignettes above, time and space are stretched and squashed in perfect tandem. Little wonder, then, that they should be connected, that they should be part of something greater. It was the Lithuanian-Polish Hermann Minkowski who was so inspired by Einstein's ideas that he made the first leap into spacetime. 'Henceforth,' he declared, 'space by itself and time by itself have vanished into the merest shadows and only a kind of blend of the two exists in its own right.' Rather wonderfully, Minkowski had once taught the young Einstein at the

Federal Institute of Technology in Zurich, although he remembered him as a 'lazy dog' who was 'never bothered about mathematics'.

What did Minkowski really mean by spacetime? To understand this, we must begin with the three dimensions of space. There are three dimensions because you need to list three independent coordinates to specify your spatial location: think of your two GPS coordinates, alongside your height above sea level. Now take a look at your watch and make a note of the time. Pause for 30 seconds and look at your watch again. Those two moments where you looked at your watch occurred at the same point in space but at different points in time. We could distinguish them by allocating a time coordinate to represent the moment at which each particular event happened. Thus, we have a fourth independent coordinate – a fourth dimension. Put them all together and we have spacetime.

To properly appreciate the elegance of spacetime we should think about how we measure distances, first in space and then in spacetime. Distances in space can be measured using the Pythagoras theorem. You probably remember this as the high-school verse about right-angled triangles – the square of the hypotenuse is equal to the sum of the squares on the other two sides – but there is much more to this ancient theorem than you might have originally thought. To appreciate why, we first set up a pair of perpendicular axes, as shown in the left-hand figure below.

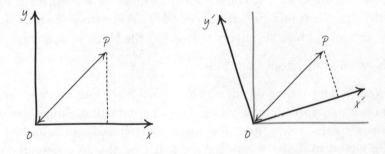

With respect to these axes, the point P has coordinates (x, y) and, by Pythagoras, is easily seen to lie at a distance $d = \sqrt{x^2 + y^2}$ from the origin. If we rotate the axes about the origin O, as shown in the right-hand figure, and define a new set of coordinates (x', y'), the distance

from the origin obviously remains unchanged and Pythagoras's theorem works just as well as before:

$$d^2 = x^2 + y^2 = x'^2 + y'^2$$

This is the real beauty of Pythagoras: its ability to remain unchanged
even when you rotate the coordinates.

Now for spacetime. Minkowski told us to mash space and time
together. Of course, we really want to mash three dimensions of
space together with our single time dimension, but to keep things a
little simpler let's just take one space dimension, labelled by the coordinate x and put that together with time, labelled by the coordinate t.
To measure distances, d, in this spacetime, Minkowski reckoned we
should use a weird form of Pythagoras, given by

$$d^2 = c^2t^2 - x^2$$

I know: the minus sign. What is that all about? We'll come to that,
but first we need to understand the c^2t^2 bit. We want to measure distances and, to state the obvious, time is not a distance. To turn it into
a distance we need to multiply it by a speed, and what better to use
than the speed of light? This means that c^2t^2 can be read as units of
distance squared, which is exactly what we want when thinking about
Pythagoras. Now for the minus sign. The spacetime measure of distance ought to remain unchanged whenever we perform the analogue
of a spacetime rotation: that is, the transformations that take us
between observers moving relative to one another, such as the one
that took us from Usain Bolt's parents to Usain Bolt himself. These
'rotations' are officially known as *Lorentz transformations*, encoding
all the stretching of time and squashing of space that makes the physics of relativity so wonderfully bizarre. The mysterious minus sign is
crucial for keeping the spacetime distances unchanged whenever you
perform this switch between inertial observers in relative motion. Perhaps this is easiest to see for light, which is travelling through space at
speed $x/t = c$. Plugging this into Minkowski's formulae,[4] we see that light
is at a *vanishing* spacetime distance from the origin. The origin stays put
whenever we 'rotate' our spacetime coordinates, so light must look

the same for all observers. Nothing moves faster than light in space, but in spacetime light doesn't move any distance at all. That's what makes it special.

What about you? What are you doing in spacetime? Well, I assume you are sitting comfortably in a chair reading this book. Whatever you are doing, we know that you are not moving in space defined with respect to yourself, but you are moving in time, so you must be moving in spacetime. How fast are you moving? Well, using the spacetime measure of distance with $x = 0$, we get $d = \sqrt{c^2 t^2}$ and so it is easy to see that you are moving through spacetime at a speed $d/t = c$. In other words, you are moving through spacetime at the speed of light. So is everyone else.

By combining his spacetime coordinates with a measure of space-time distance, Minkowski was starting to build a remarkably elegant picture of physics in terms of four-dimensional geometry. When Maxwell's equations are written in this new language they take on an incredibly simple form. Keeping space and time separate is like staring at the world through a fog. Bring them together and a world of remarkable beauty and simplicity is revealed. That's what makes theoretical physics such a wonderful thing to study: the more you understand, the simpler it gets. Perhaps this was no more apparent than when Einstein used geometry to conquer the gravitational force, to see that gravity is fake. That story will come next, told, as ever, through the slowing of time. But we won't be running alongside Usain Bolt or hurtling through space with Gennady Padalka. We'll be plunging towards the centre of the Earth, where time ticks a little more slowly than it does at the surface.

THE CHALLENGER DEEP

'It's really the sense of isolation, more than anything, realizing how tiny you are down in this big, vast, black, unknown and unexplored place.'

These were the words of Canadian film director James Cameron. They betray a palpable sense of fear, of no longer being in control, of being at the mercy of something greater. They would not be out of place in the script of his most famous movie, *Titanic*, but instead they expressed his emotions upon his return from the Challenger Deep, at the bottom of the Mariana Trench, the deepest known point on the Earth's seabed, almost 11 kilometres below sea level. On 26 March 2012 Cameron journeyed there aboard the deep-sea submersible known as the *Deepsea Challenger* and spent three hours exploring this alien world, all alone in the most hostile environment on the planet.

Cameron was the first person to plunge to such remarkable depths since a US naval team fifty years earlier, and was the first to do so alone. Perhaps the most remarkable fact of all, however, is that he returned from his trip having leapt forward in time by 13 nanoseconds.

Cameron's leap into the future was not due to his high speed, as with Usain Bolt or Gennady Padalka, but due to his *depth*. You see, time also slows as you plunge deeper into a gravitational well; in this case, as you plunge closer to the centre of the Earth. This is an effect of the general theory of relativity – relativity combined with gravity, and the zenith of Einstein's genius. Because James Cameron spent so long exploring the deep, he accumulated an impressive amount of gravitational time dilation. That said, it was the crew of the Arktika 2007 expedition who went closer than any other to the centre of the Earth. On 2 August 2007, pilot Anatoly Sagalevich, polar explorer Artur Chilingarov and businessman Vladimir Gruzdev were the first to descend to the Arctic seabed aboard *MIR-1*, some 4,261 metres below the surface at the North Pole. This might not seem like much compared to the depth of the Mariana Trench, but the Earth is not a perfect sphere. It is an oblate spheroid, bulging out slightly at the equator. As a result, the crew came much closer to its centre than *Deepsea Challenger*. After an hour and a half on the seabed the three men on board *MIR-1* had skipped forward in time by a few nanoseconds. As well as taking soil and animal samples, they planted a Russian flag made of rust-proof titanium metal. The incident sparked fierce objections from other Arctic nations, who saw it as a move to claim the region as Russian territory. The Russians denied this, stating that their goal was simply to prove that the Russian shelf extended as far as the North Pole and comparing it to the moment the *Apollo 11* astronauts planted the American flag on the surface of the Moon.

Although this is not a book about international politics, in this part of the story such things are never too far away. To understand how and why these deep-sea explorers were able to slow down time, we need to position ourselves in the early part of the twentieth century, at a time when the world was at war, the trenches filled with the blood of ordinary men fighting in extraordinary circumstances. At this time there was also a battle raging in the world of science. British physics had been reluctant to embrace Einstein's new ideas about time and space. More than any other community, the British were still invested in the notion of the aether, led, no doubt, by the indomitable Scots-Irish baron Lord Kelvin. They were also invested in Isaac Newton, the legend of British science, whose laws of universal gravitation were still the established model some three hundred years after they were first proposed. Newtonian gravity could explain so much: from the motion of the planets to the trajectory of bullets raining down at the battle of the Somme. But there was also something troubling about Newton's theory, something that Einstein's work brought into sharper focus: instantaneous action at a distance.

To understand why, imagine what would happen if the Sun were to spontaneously disappear in an instant. Of course, we would all die, but how long would it take for us to become aware of our fate? In a world ruled by Newtonian theory, the force of gravity acts instantaneously over large distances, so we would know about the Sun's demise the moment it happened. The trouble is that it takes eight minutes for sunlight to reach us here on Earth. From Einstein's perspective, this means that it should take at least eight minutes for us to receive *any* signal from the Sun, including one that alluded to its demise. Clearly Newton and Einstein are in direct conflict. Although Einstein was far from patriotic, a German challenge to the Newtonian throne was never going to be well received in England against the backdrop of the Great War.

Newton himself had serious misgivings about this action at a distance. In a letter to the scholar Richard Bentley in February 1692 he wrote, 'that . . . one body may act upon another at a distance through a vacuum wthout [sic] the mediation of any thing [sic] else . . . is to me such an absurdity that I beleive [sic] no man who has in philosophical matters any competent faculty of thinking can ever fall into it'.

Einstein would eventually address these concerns, but to do so he would deny Newton and refute his greatest discovery. He would deny the existence of gravity altogether.

Gravity is fake.

I like to start my Advanced Gravity class with this little one-liner, even though it upsets some of the students. But the statement *is* true: gravity really is a fake. Even on Earth, you can become weightless; you can eliminate gravity altogether. To see how, take a trip to the opulent desert city of Dubai and climb to the top of the Burj Khalifa, the world's tallest building, stretching almost a kilometre up into the sky. Once there, get inside a large box, something like an old British telephone box with the windows blacked out, and have someone drop you over the edge. As you fall with the box towards the ground, what will happen? You are accelerating towards the Earth at 1g, but *so is the floor of the box*. OK, so there is a small amount of air resistance that will drag on the box, but if the air is thin enough, you will more or less become weightless and gravity will disappear. Now, I appreciate that this is a drastic way to test gravity. But actually, you don't really need to jump off the Burj Khalifa to feel the effects of weightlessness. It is enough to drive down a steep hill in your car. You probably already know that feeling as your stomach starts to perform somersaults. That is gravity starting to disappear as you accelerate down the hill. Whenever it happens, I always remind myself (and anyone who is in the car with me), that they are feeling the effects of Einstein's genius right there in their belly.

When Einstein saw that he could always eliminate the effects of gravity, he declared it to be the happiest thought of his life. The death of gravity can be traced all the way to Galileo, the genius of the Renaissance and the founder of modern science. According to his student Vincenzo Viviani, Galileo would drop spherical objects of different mass from the top of the Leaning Tower of Pisa, demonstrating to the professors and students how they fell at the same rate. This contradicted Aristotle's ancient claim that heavier objects would fall faster. Whether or not Galileo ever really put on such performances is a matter of some debate,* but the effect is certainly real. A version of

---

* Most scholars believe that he only ever performed it as a thought experiment, although Canadian historian Stillman Drake has argued that Viviani's account was broadly accurate.

his experiment was even carried out on the Moon, by *Apollo 15* astronaut David Scott. He held a hammer in one hand and a feather in the other then simultaneously dropped them towards the lunar surface. Without air resistance, the two objects fell at exactly the same rate, just as Galileo had predicted. It is precisely this universal behaviour that guarantees that both you and the telephone box fall from the Burj Khalifa in perfect tandem.

If we can eliminate gravity altogether, in what sense is it real? Can we fake it in outer space? Faking gravity in space is easy – all you need to do is accelerate. If the International Space Station were to switch on its boosters and begin accelerating towards higher altitude at 1g, the astronauts would immediately cease to feel weightless. The ship would push upwards, but to the astronauts it would feel as if they were falling down, just as they would under the influence of gravity. Black out the windows and they could well be fooled into thinking that the ISS had come crashing down to Earth.

The point here is that gravity and acceleration are indistinguishable – in a blacked-out spaceship you have no way of knowing if you are feeling the effects of gravity or if the ship is accelerating through space. This is known as Einstein's *equivalence principle* – the physical equivalence between gravity on the one hand and acceleration on the other. You cannot tell the two of them apart. If you are still not convinced, think about what happens when you are driving your car and you take the corner a little too quickly. Turn left and it's as if you are pulled towards the car door on the right. This is just like a fake force of gravity acting sideways. The truth is that it is the car that is accelerating as it turns the corner while your body wants to carry on in the same direction, the result being that you swing towards the opposite car door.

Let's return to our deep-sea explorers for a moment. To fully appreciate how time is slowed down for them we need to think about light again. How does gravity affect light? Since gravity and acceleration are indistinguishable, we may as well just ask how acceleration affects light. Imagine that you are in a spaceship cruising through empty interstellar space at constant speed and resting in your arms is a plate of jelly.[5] In contrast, your friend is carrying a laser gun. If this were a duel, you would lose, but it's not, it's an experiment. You tell your friend to fire the laser at the jelly. She does as you ask and the laser slices through the jelly in a perfectly straight line. You decide to try again, only this time you fire the engines and start to accelerate the rocket. You and your friend immediately feel the effect of the fake gravity and are able to stand as normal on the floor of the spaceship as it pushes you through space. You tell her to fire the laser, which she does, and again the jelly is sliced through. You take a closer look at the paths that the laser made. While the first path went straight through the jelly, the second is slightly arced, as shown below.

What happens when you fire a laser at a plate of jelly in space, if the spacecraft is travelling at constant speed (left) and if it is accelerating (right).

What has happened to the second light ray? Nothing special. It still fired through space in a straight line, as it should, but it did so while the jelly was accelerating 'upwards' with the rocket. From the point of view of you and the jelly, it is as if the light ray is bent. While this is clearly just a consequence of the jelly's acceleration, the equivalence principle suggests that light should also be bent by gravity.

And it is.

The proof arrived not long after the Great War ended. Although few people had embraced Einstein's new ideas in Britain during this difficult time, he did have one advocate. Arthur Eddington was a thoughtful and ambitious astronomer, a pacifist who encouraged British scientists to maintain their pre-war interest in the work of German colleagues. Though it was hard to access German scientific journals, he knew about Einstein's work through the Dutch physicist Willem de Sitter and was determined to test the prediction that starlight would be bent by the Sun's gravity. The trouble with observing starlight passing close to the Sun is that the Sun's glare makes it impossible to see. Eddington realized that he needed a solar eclipse to perform the experiment and his calculations suggested that one was due to take place on 29 May 1919 on the beautiful Portuguese island of São Tomé and Príncipe, off the west coast of Africa, before moving across the Atlantic to northern Brazil. Eddington travelled to the African island with Astronomer Royal Frank Watson Dyson, while a second team was dispatched to observe the eclipse from Sobral in the Brazilian state of Ceará. Despite cloud and rain threatening the success of the experiment, the team were able to photograph several stars in the Hyades cluster during the eclipse. When these were compared to night-time images of the same cluster, the images did not align. The implication was that the starlight passing closest to the Sun had been bent more in the eclipse photograph, creating a mismatch with the night-time images. Einstein's prediction was confirmed and made headline news across the globe. It was the moment he became a superstar.

The bending of light has important implications for time. Far away from a gravitational field, when light is travelling in a straight line, it takes just a few nanoseconds to travel from a lamp on one wall of the ISS to a picture on the other. But if we placed the ISS in orbit around a black hole, this light would be bent by the strong gravitational field. Curved paths are longer than straight ones, so the light would take a little longer to complete its journey from one wall to the other. This means that the same event takes longer to happen when there is more gravity, and so gravity must be slowing down time.

The stronger the gravitational field, the more the light will be bent, and the more time will slow down. This is why James Cameron was able to leap into the future by diving to the bottom of the Mariana Trench. The gravitational field of the Earth is stronger there, albeit by a tiny amount, so clocks tick more slowly. The reverse is also true. Climb high and the gravitational field will weaken slightly, causing clocks to tick more quickly. A second spent at the summit of Mount Everest is about a trillionth of a second longer than the amount of time spent at sea level. After their twelve-and-a-half-day mission, including three days on the Moon, the *Apollo 17* astronauts experienced a record negative time dilation, going back in time by around a millisecond.*

---

* The crew of *Apollo 17* were travelling at high speeds for much of their journey, which would have had the effect of slowing down their time, but for most of the mission the

The effects of gravity on time were measured *directly* in a famous experiment that took place at the Jefferson Tower at Harvard University in 1959. Robert Pound and his student Glen Rebka Jr fired gamma rays – high-energy electromagnetic waves – from the top of the 22.6-metre-tall tower to a receiver at the bottom. Their clever idea was to use the frequency of the gamma rays as a measure of time, the clock 'ticking' with each new oscillation of the wave. As it turned out, the *same* waves were measured to have higher frequencies at the bottom of the tower than they did at the top. That meant that a single second at the bottom corresponded to more oscillations of the wave than a second at the top. There was only one conclusion – the meaning of 'a second' had to be different at the two ends of the tower. A second at the bottom represented more oscillations of the wave, so it must have been a *longer* second. Time was ticking more slowly at the bottom of the tower than it was at the top, just as Einstein had predicted.

Gravity's ability to bend light and slow time means that the Earth's core is about two and a half years younger than its surface.[6] But how does gravity do this if it's really a fake? How does it cause the bending of light? The truth is that it doesn't. Light always travels in a straight line through space – it is the space itself that is bent. To picture what is going on, go to the fruit bowl and grab an orange. Mark two points on the surface of the orange, reasonably far apart, and then draw the shortest path between the points. If you aren't quite sure which is the shortest path, line the points up so that they are both level at the same height on the orange's 'equator', then draw the line along the equator. Now peel the orange carefully so that the skin remains in one piece. When you have done this, flatten out the peel on the table. What shape is the line that you drew? It is curved, right? This is very weird, because the shortest distance between two points is supposed to be a straight line, but it turns out that this is only true on a flat surface. On a curved surface, the shortest paths are curved, just like the one you drew on the orange. That is what light is doing. It follows the shortest path through space, but because space is curved the path is curved. If you've ever flown long distance from London to New York and sat

---

negative effects of gravitational time dilation at high altitude dominated over the special relativistic effect.

there watching the flight map, you will have noticed how the aeroplane always looks to be taking a strange, curved trajectory up through the Canadian Arctic. This is because the airline has calculated the shortest path, and it is curved, just like the surface of the Earth.

Of course, it is really the spacetime geometry that is curved. Minkowski told us how to measure distances in a flat spacetime geometry, but when it becomes curved the distance measures get squashed and squeezed, stretched and pulled. What causes this squashing and squeezing? Matter. You. The Sun. The Earth. Anything with mass, energy or momentum causes spacetime to bend and warp. Imagine a rubber sheet stretched out flat. Throw a heavy rock on to the sheet and it causes it to curve. That's a good analogy for what matter does to spacetime.

Light will follow the shortest paths on this curved spacetime. It follows a very special kind of shortest path, so short in fact that its spacetime length vanishes. But that's what makes light special, remember, and it remains true when the spacetime becomes curved. These light-like paths are called *null geodesics*. What about heavier stuff, like planets or suns? What do they do in spacetime? Well, they also follow the shortest paths available to them, the analogue of straight lines. They don't follow the same paths as light rays because they don't travel quite that quickly, but they do take the most economic route through spacetime that is available to them. These paths are known as *timelike geodesics*. In curved spacetime they are curved. In fact, they can appear very curved indeed. The Earth's path is so curved that it loops right back on to itself, mapping out an ellipse on its annual journey around the Sun. In reality, it is following a timelike geodesic, a straight line through the highly curved spacetime created by the gravitating Sun.

You may think I am using too much poetic licence in describing these curved paths as straight lines when they are very obviously not straight. Actually, I am being more literal than you probably think. It turns out that the kind of spacetime geometries we are interested in always look flat when you zoom right in. It's a bit like how the surface of the Earth looks curved when you look at it from space but, close up, on land, you might be tricked into thinking it is flat. To a good approximation it is flat, of course, as long as you stay zoomed in, and it's the same with spacetime. Zoom in close enough to even the most curved of geometries and it will look just like the spacetime Minkowski described. It is because of this ability to zoom right in and discover Minkowski spacetime that we are able to do away with gravity, at least in a small enough environment. That is what was happening when you jumped off the Burj Khalifa. Sure, the Earth sets up a curved spacetime, but jump off the world's tallest building in a telephone box and you can find yourself zooming right in and doing away with gravity altogether, at least to a very good approximation.

These shortest paths – these timelike geodesics – are the same whoever or whatever happens to be following them. Hammer or feather, it makes no difference; both will follow a timelike geodesic and travel through spacetime at the speed of light. Both objects fall in exactly the same way – just as Galileo said they would. But it took Einstein to explain *why* this happens.

Einstein's theory has triumphed time and again, its outlandish predictions confirmed by even more outlandish experiments: from the bending of light and Eddington's ambitious post-war expedition to the island of São Tomé and Príncipe, to the gravitational slowing of time and the bouncing gamma rays of Pound and Rebka. Planetary orbits provide another key test of Einstein's theory, most notably in the case of the trajectory of the planet Mercury. Although the orbit is elliptical, the ellipse itself moves, it *precesses*, adjusting its position year on year by a tiny amount. This Mercurian wobble is expected even in Newtonian gravity from the gravitational effects of the other planets, although the numbers are off. When the French mathematician Urbain Le Verrier noticed as much, he predicted the existence of *Vulcan*, an unseen dark planet lying between Mercury and the Sun. According to Le Verrier, Vulcan's gravity would be enough to give Mercury's orbit the kick it needed to wobble in just the right way. Le Verrier had built a career on this sort of prediction. In August 1846 he had predicted the existence of the planet Neptune by examining the wobbles in the orbit of Uranus.* Within a month, two German

* The Cornish mathematician John Couch Adams had independently drawn identical conclusions. However, he posted his results to the Royal Observatory in Greenwich *two days* after Le Verrier had announced his predicted location of the new planet to the

astronomers, Galle and d'Arrest, had found Neptune to be within one degree of Le Verrier's predicted location. In contrast, Vulcan was never found, despite a number of false alarms. The truth is that Vulcan does not exist and that Mercury's wobble can be accounted for with the corrections coming from Einstein's theory. Mercury feels these corrections more than the other planets because it is closest to the Sun.

This cautionary tale, the contrasting fortunes of Neptune and Vulcan, echoes through to the twenty-first century. Today we argue about the need for *dark* matter and *dark* energy to bring our theory in line with cosmological observations. It has been suggested that these are no more real than Vulcan and that what we are seeing are the corrections from an even newer theory of gravity, an improvement on Einstein's theory that is relevant to astrophysics and cosmology. While this idea gained some momentum around the turn of the millennium, it has stalled recently after yet another boon for Einstein's original theory: the discovery of gravitational waves in 2015. Einstein predicted that spacetime was a dynamical beast, that it should contain ripples, waves of gravity flowing through it, distorting the shape of time and space in a very particular way. Alternative theories often predict waves that distort spacetime differently, but the ones we've measured match perfectly with Einstein's original prediction. It would be a gravitational wave or, perhaps more accurately, a spacetime tsunami, that would alert us to the Sun's demise if it were to miraculously disappear. The wave would travel across the solar system at the speed of light, tearing up the gravitational field of the Sun, a final apocalyptic verification of Einstein's triumph over Newton.

If Usain Bolt was the limit of human relativity, the zenith of our physical ability to meddle with time, then what is gravity's equivalent? Where does gravity distort time beyond all recognition? The answer lies in 'an embellished dark source of unending creation'.

It lies within Pōwehi.

French Academy. In keeping with his Cornish roots, Adams had taken things slowly, starting his calculation before Le Verrier but finishing slightly later.

A GLIMPSE INTO THE ABYSS

Pōwehi. The word is Hawaiian, taken from the *Kumilipo*, an ancient chant describing the creation of the universe, the 'embellished dark source of unending creation'. In Maori, it simply means horror. Pōwehi is a monster, a terrifying behemoth, lurking at the core of Messier 87, a supergiant galaxy in the constellation of Virgo. In April 2019, those of us on Earth beheld it for the first time.

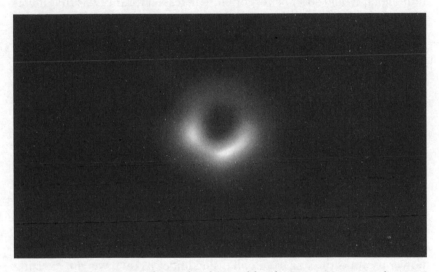

The spectacular image of Pōwehi observed by the Event Horizon telescope.

The startling image of Pōwehi was captured by the Event Horizon telescope, an array of eight ground-based radio observatories strategically positioned across the globe. It was an extraordinary accomplishment, given the size and distance to the source. Imagine yourself sitting in a Parisian café and peering through your telescope to read a newspaper in New York. That is what it took to capture this astonishing image in such magnificent detail.

But what is it, this horror, this dark source? Pōwehi is a black hole of gargantuan proportions, billions of times more massive than the Sun. It is gravity taken to its terrifying limit. We have already seen how light is bent by gravity. What happens as you ramp up the gravitational field, as you curve the spacetime more and more? You create a prison. Light is bent to such an extent that it becomes trapped, it cannot escape, and if light cannot escape, nothing can. Pōwehi is a cosmic oubliette, an unforgiving hell, a gaol for the forgotten.

It was an English clergyman who first conceived of such horrors. In November 1783 the Revd John Michell proposed the existence of *dark stars*, huge astrophysical objects five hundred times larger than the Sun whose gravitational pull was so strong that light itself could not escape.* It was an exciting idea at the time, invisible giants hiding in plain sight, although it would soon be forgotten. The reason for this was that it was based on the corpuscular theory, where light is made up of particles, a theory that ultimately gave way to a wave-like model following the experiments of Thomas Young at the turn of the nineteenth century. Although Michell's work on black holes would be ignored for almost two centuries, he would be heralded in science as the father of seismology. His work on the devastating earthquake and tsunami that struck Lisbon in 1755 included the idea that it originated from faults in the Earth's crust rather than from atmospheric disturbances.

Today most scientists are confident that black holes really do exist. Typically, they form when a sufficiently large star – at least twenty times heavier than the Sun – runs out of fuel. Stars power themselves with nuclear fusion, squashing and squeezing atomic nuclei together in their core, a furnace of thermonuclear bombs exploding continuously. This power prevents the star from collapsing under its own weight, exerting outward thermal pressure to counter the effects of gravity. But it doesn't last for ever. Once the star has produced too much iron in its core, the fusion processes become inefficient and it can no longer support its own weight. Star death. Gravity quickly begins to overwhelm the star, crushing it inwards, a garrotte that gets tighter and tighter. And then *bang*! The star fights back, a dramatic counterpunch to gravity's relentless attack. It is the neutrons that carry the fight, subatomic

* The brilliant French mathematician Pierre Simon Laplace drew similar conclusions about the possibility of black hole-like objects a decade or so after Michell. It is not clear to what extent he knew of Michell's work. France was in the grip of revolution at this time so scientific communication between their two countries would certainly not have been easy.

particles in the stellar core, violently repelling one another through a strong nuclear force whenever they are pushed too close together. Outer layers of material fall inwards, strike the immovable core of neutrons and rebound. In an instant, a pressure wave powers its way to the surface of the star and it explodes. A supernova, a cataclysmic event, briefly outshining an entire galaxy.

What is left behind? More than likely a neutron star, an object of tremendous density, so much so that a mere teaspoonful of its matter would weigh as much as a mountain here on Earth. If its total mass can stay below that of about three Suns, the neutron star has a chance of survival. Any heavier, and the gravitational garrotte will begin to tighten once more. There will be nothing the neutrons can do. There will be nothing anything can do. The collapse becomes unstoppable. Eventually, the star becomes so dense that light can no longer escape. Everything that was once the star is hidden behind an *event horizon*, the trapdoor to the cosmic oubliette, a spheroidal surface beyond which there is no return.

About one in every thousand stars is heavy enough to end its life consumed by gravity. These *stellar mass* black holes are everywhere, scattered across the galaxy, shadowy remnants of the largest and most powerful stars ever to have existed. But Pōwehi is so much more. Black holes born from star death typically weigh between five and ten Suns and yet Pōwehi has a mass of six and a half *billion* Suns. A leviathan, a supermassive black hole, the anchor at the core of an enormous galaxy more than 50 million light years away. Pōwehi dwarfs our own leviathan, Sagittarius A*, a black hole of 4 million solar masses at the centre of the Milky Way. Most galaxies are thought to be anchored around a supermassive black hole. Galaxy 0402+379 contains two such leviathans, probably as a result of two daughter galaxies colliding. The core of 0402+379 must be a raging tsunami of gravitational waves, tearing through spacetime as the two leviathans wrestle for supremacy. The truth is that we don't fully understand how Pōwehi or any of these other monsters came to be. It is possible they are the greedy remnants of giant stars, once stellar mass black holes that grew to gargantuan sizes after millions of years of feeding on any material that dared to stray too close.

The existence of the event horizon *defines* the black hole. Just to stay still on its surface you would need to travel at the speed of light. For a stellar mass black hole, edging close to the horizon would be fatal. In a way, this is weird; gravity is fake, remember, and we can always eliminate it by climbing inside the blacked-out telephone box and falling, be it from the Burj Khalifa or towards the event horizon of a black hole. The trouble is that the region over which we can eliminate it – the size of the telephone box – gets smaller and smaller as the gravitational field grows stronger, as the spacetime becomes more strongly curved. Beyond the box there are dangerously large gradients in the gravitational stress, tides of gravity that cannot be ignored. For a stellar mass black hole, the horizon is too close to the bottom of the well and the tides of gravity would tear you apart as soon as you got too close. On the other hand, for a supergiant black hole like Pōwehi, the bottom of the well is further away so passing through the horizon is unremarkable. Once you have crossed this threshold, however, your days are numbered. Literally. Time will end. At the core of the black hole is a *singularity*, a place where spacetime touches infinity, where the gravitational field grows without bound. The singularity is not an end of space but an end of time. Once you cross the event horizon, your trajectory through spacetime will take you there, to a place where there is literally no tomorrow, where the future does not exist – not even in principle. As you approach this Armageddon, the gravitational stresses, those monstrous tides, stretch you out like a string of spaghetti, the atoms in your body torn apart, the nuclei ripped into protons and neutrons, the protons and neutrons ripped into their constituent quarks and gluons. Whatever consciousness is left will seek the end, and the end will come at the singularity, a merciful inevitability.

However, if others were to watch you fall into the black hole from afar, they would see a very different picture. At first, they would see you accelerate towards oblivion, and if they could somehow see your subjective clock, the watch on your wrist, they would see it slow more and more as you plunged deeper and deeper into the gravitational well. As you approached the threshold, it – and you – would appear to slow to a complete halt. It would be as if you were frozen in time and space, decorating the horizon with a permanent reminder of what can happen when you stray too close. It is not that you didn't cross into the black hole; you did, it's just that those outside could never see you do it because every second you experienced at the horizon would be an *eternity* to them.

For objects away from the horizon, time will not stop, but it will slow down considerably if they get too close. If the black hole has enough spin, there can be *stable* planetary orbits that veer very close to the horizon and, in principle, you could visit these for a while, slow down time and then return home catapulted years into the future. In the film *Interstellar*, the crew of the *Endurance* experience the full force of gravitational time dilation by visiting Miller's planet, orbiting a supermassive black hole called Gargantua. Gargantua is assumed to be spinning so fast – within a trillionth of a per cent of the theoretical maximum – that Miller's planet can orbit within a few thousandths of a per cent of the horizon radius.[7] The reconnaissance crew visit the planet for a little over three hours, yet they return to find their colleague, who had stayed aboard *Endurance*, aged by a staggering twenty-three years. That said, black holes with this amount of spin will be incredibly rare, if they exist at all, since there are natural mechanisms to prevent the spin from increasing beyond 99.8 per cent of the maximum. This means the planetary orbits cannot edge *quite* so close to the horizon and the dilation effects are weaker. The spin of Pōwehi could well be around this 99.8 per cent mark. Three hours or so on an innermost planet orbiting this real-life leviathan would then equate to thirty-two hours and twenty-four minutes for those waiting on the mothership. Although this is not quite Hollywood, we should remember that Pōwehi is *real*, we have seen it, and perhaps some of its planets are inhabited by beings whose lives tick along almost eleven times more slowly, in comparison to our frenzied existence here on Earth.

The image of Pōwehi is compelling evidence for the existence of black holes in Nature – make no mistake about that – but it is not conclusive. After all, we do not see the event horizon itself, but a shadow that is two and a half times larger. Despite the remarkable and inspiring imagery offered by the Event Horizon telescope, the strongest evidence for black holes comes from gravitational waves. On 14 September 2015 the team at LIGO, the Laser Interferometer Gravitational Wave Observatory, detected these tiny ripples in the fabric of spacetime for the very first time. LIGO operates across two sites: one in Hanford, Washington – a decommissioned nuclear production complex – and the other in the alligator-infested swamps of Livingston, Louisiana. These ripples were tiny, stretching and squeezing the 4-kilometre arms of the detectors by less than the width of a proton, betraying their violent beginnings from the merger of two black holes, the mass of thirty-six and twenty-nine Suns respectively, in the furthest reaches of the observable universe. The energy carried by the wave at the source was spectacular, equivalent to the mass of three Suns, or 10^{34} Hiroshima bombs, an explosive spacetime tsunami crushing space one way and stretching it the other. But could it have been something else that generated the wave, a coming together of some other exotic compact object different from a black hole? At the point of their merger, the two objects were just 350 kilometres apart, a combined mass of sixty-five Suns crammed into a region less than twice the size of the would-be event horizon. It's hard to imagine it was anything other than a pair of black holes spiralling towards the ultimate embrace.

1.00000000000000858 didn't seem like a big number in the beginning, but it was big enough to open the door to an unfamiliar world. When Usain Bolt powered his way to this world-record time dilation, he touched the edge of relativity. He encouraged us to glimpse into a world of physics removed from everyday intuition, where running tracks shrink and time is slowed down. At its most extreme, this is the physics of black holes, where time is brought to a standstill for the wretched victim who falls into the horizon. We are lucky enough to live in an unprecedented era for black-hole discovery: we can see the dark shadow cast by the giant Pōwehi at the heart of a monstrous galaxy; we can listen to leviathans collide through gravitational waves that roar across spacetime like a relativistic clap of thunder signalling the marriage of celestial gods. The physics of these gods suggests a *shadowy* truth about our physical reality – a holographic truth, a universe trapped in a hologram. It is a tale we will continue to tell in the forthcoming chapters, as we explore ideas about entropy, the guardian of secrets, and quantum mechanics, sovereign to a subatomic world. It is a tale to be told through leviathans, numbers that are bigger and even more remarkable than 1.00000000000000858.

A Googol

THE TALES OF GERARD GRANT

My cousin, Gerard Grant, liked to tell us ghost stories when we were kids. He told us about the time he saw the ghost of his grandfather, lit by moonlight, praying before a statue of the Virgin Mary; or the time when he was camping in a remote part of Ireland and woke to discover bacon and eggs sizzling on the stove outside his tent. It must have been the 'little people' he said, 'the leprechauns.' There was even the tale of the man who predicted his own death. 'He saw himself walking behind himself,' Gerard told us, with a sense of foreboding. 'His doppelgänger. His exact copy. He knew then he was going to die.' And then he did, or so Gerard said.

You might think that stories about doppelgängers have no place in a serious book about physics and mathematics. But you should expect the unexpected when it comes to tales of numerical gargantua. This particular tale begins with the *googol*:

10,000,000,000,000,000,000,000,000,000,000,000,000,000,000,000, 000,000,000,000,000,000,000,000,000,000,000,000,000,000,000, 000,000

That's a one followed by a *hundred* zeroes, or ten to the power of one hundred. There is a decimalized elegance to the googol, perhaps even a decadence, and we can safely describe it as a big number by any earthly standard. If you won a googol pounds on the lottery, you could buy yourself a luxury yacht, or even a fleet of luxury yachts, an aircraft carrier, or, if you prefer, every shipping vessel on the entire planet. You could even buy the United States of America. The USA, in its entirety, would probably cost you less than 50 trillion dollars, which would be nothing to a googolaire like yourself. You could literally buy everything: every molecule, every atom, every fundamental particle in the observable universe. There are around 10^{80} particles in the universe, so you could afford to do this and still pay more than a quintillion pounds for each one.

The legend of the googol really begins with Milton Sirotta, a nine-year-old boy whose Uncle Edward just so happened to be the eminent mathematician Edward Kasner of Columbia University. Kasner belongs to a select group of people who have their very own space-time, such as Hermann Minkowski, Karl Schwarzschild and Roy Kerr. The Kasner spacetime is unlike any universe you have ever experienced. If you were sitting inside it, you would find some directions of space expanding and others contracting, like a piece of dough being stretched one way and squashed in the other. But this terrifying world has nothing to do with a googol. When he came up with that particular concept, Kasner was trying to convey the vastness of infinity. His point was that even those numbers that seemed very large were in any practical sense vanishingly small when compared to the infinite. He chose to illustrate this fact using a one followed by a hundred zeroes but needed a name for this tiny behemoth. Ten duotrigintillion or 10 sexdecilliard really would not do. His nephew, Milton, had a much better suggestion: the googol.

It is amusing to note that a number lauded for being so large was originally introduced to demonstrate how small it was. Kasner and his nephew soon came up with another fantastic number: the *googolplex*. By Milton's original definition, a googolplex was a one followed by zeroes 'until you get tired'. To find out how large that would actually be I did an experiment: in one minute I was able to write out a one followed by 135 zeroes at a fairly relaxed pace without getting tired, so a googolplex is certainly larger than a googol. To push things a little harder, we would do well to enlist someone with the stamina of Randy Gardner. As a teenager in the mid-1960s, Randy stayed awake for eleven days and twenty-five minutes as part of an experiment on the effects of sleep deprivation. If he had spent that time writing out a googolplex, working continuously at my leisurely writing pace, he would have written out a one followed by 2,141,775 zeroes. That's big, but ultimately Kasner demanded a less vague definition of the googolplex and settled on a number far beyond the reach of Milton's criteria. Kasner defined his new number to be a one followed by a googol zeroes. Breathe that in: a googol zeroes! Ten to the googol! While this seems truly colossal, Kasner's point was that there are an infinity of numbers that are far larger.

Like the googolplexian. That's a one followed by a googolplex zeroes. A googolplexian is also referred to as a *googolplexplex* or a *googolduplex*. In fact, these latter definitions are much more powerful since they allow us to exploit the idea of recursion to generate a whole tower of truly enormous numbers. From a googolduplex you can leap to a *googoltriplex*, a one followed by a googolduplex zeroes, and then to a *googolquadruplex*, which has a googoltriplex zeroes, and so on.[1]

But we are getting carried away. We shall pause at a googol and a googolplex because they are already more than enough to elucidate the next piece of remarkable physics and to revisit the cautionary tale of the doppelgänger. You see, when we imagine a googolian, or even a googolplician universe, we can start to ask if *doppelgängers are real*. By a googolian universe, I mean one that is at least a googol across in any earthly units of distance you choose to use (metres, inches, furlongs – it really doesn't make that much difference). A googolplician universe is even bigger, a googolplex across in the same earthly units.

The idea of cosmological doppelgängers goes back to MIT physicist Max Tegmark.[2] He imagined a vast and magnificent universe with distant worlds far beyond the reach of any telescope, and estimated the distance to your exact copy somewhere far away in the cosmos, with the exact same style of hair, the same nose, even the same thoughts. When I first read his claim, I was sceptical. No offence, but why would the universe need another version of you? Or me? Or James Corden? And then I sat down and thought about it. Tegmark's claim was a consequence of the holographic world, the greatest illusion in all of physics.

I decided to estimate the distance for myself, using some of the most important ideas that led some of the world's greatest physicists towards a holographic truth. It's a story I will need to tell you over two chapters, between a googol and a googolplex. It begins with entropy and what it could mean for a human and a black hole of human size. It takes us deep into quantum theory, the magical ingredient of a microscopic world, establishing what it really means for you to be you and your doppelgänger to be you. In the end, my final estimate is a little more conservative than Tegmark's, but it is in the same ballpark. I reckon the distance between you and your doppelgänger, in metres, miles or whatever earthly units you care to use, is somewhere in between our next two leviathans: a googol and a googolplex. In other words, you won't find your doppelgänger in a googolian universe, but in a *googolplician* universe they will almost certainly exist. They might even be reading an exact copy of this book.

THE ENTROPIC CAPTOR

Take a look in the mirror. What do you see? Whenever I look at my own reflection, I normally notice the speckles of grey hair or the crisscross wrinkles inherited from my Spanish grandma – my *abuelita*. These things don't bother me. After all, I'm a theoretical physicist. As a profession, we are not well known for worrying about our appearance. But what I do see is the passage of time – the ascent of *entropy*.

If we are to estimate the distance to your doppelgänger, we must first understand entropy, and the terror of its ascent. Entropy is often misrepresented, carelessly used as a synonym for disorder or disruption. In fact it is better understood as a captor or a gaoler. It is the turnkey that locks energy away for good – including, one day, that of the entire universe. Imagine yourself, for a moment, in Victorian England. You are looking down on the clouds of black smoke rising from the chimneys of a northern town. There are workers filing like ants into factories, their terraced homes shrouded in a smoggy satanic haze. This was when our appetite as humans first became insatiable: more machines, more energy, more power. But it can't go on for ever – not because the planet is dying from the effects of climate change but because of entropy and its formidable ascent.

The story of entropy began in these Victorian factories and in the inquisitive mind of a young French military engineer by the name of Sadi Carnot. Inspired by the smoke and thunder of the Industrial Revolution,

Carnot invented his own branch of physics – *thermodynamics* – devoted to the dynamics of heat and its relationship with mechanical power. Whenever you burn fuel, the goal is to convert that heat into something useful. For example, in the engine of your car, fuel burns very quickly, releasing hot gases that push back the pistons. This motion is then transferred to the wheels via the crankshaft, thrusting the car forward. There weren't any cars in the early part of the nineteenth century, but Carnot's ideas applied far beyond the trains and factories of his day. He understood that the key to an engine is a difference in temperature. Whenever a difference exists, you can extract useful mechanical work like the forward motion of a train or the motive power of a machine. But heat will always move from hot to cold until the temperature difference is gone, and then it is over. You cannot extract any more work and you won't be able to power any more machines.

You might think that you can shuffle heat around, perhaps even using your machine to heat things up or cool things down again. The hope would be to re-create the temperature difference so that you can once again extract some useful work. This is true to some extent, but Carnot was able to show that shuffling things around in this way always required you to put in more useful energy than you got out. In the context of a car, the idea would be to convert the kinetic energy of the car back into fuel, saving you the hassle of going to fill up at the petrol station. If you were clever enough, you might get *some* of that energy back, but not as much as you originally put in, and eventually your engine would run dry. The problem is that in the real world you always lose a little something. You can never reset your engines completely, at least not for free. This kind of knowledge was important to the Victorian entrepreneurs thinking about potential profits from their factories. As we will see, it will also be important to us in understanding how the entropic psychopath is strangling the life out of the entire universe.

It is hard to decide what is most remarkable about Carnot's work: that he figured all of this out before anybody knew anything about energy conservation (we'll come to that in a moment), or that he did so while imagining a model of heat that just so happened to be completely wrong. Like many of his contemporaries, he believed that heat behaved like a fluid, a self-repelling substance known as caloric. Caloric does not exist. In the end, that didn't matter, thanks to Carnot's unique ability to strip away the details and focus on what was really important. Within four years of publishing his ideas Carnot had retired from the army and within a decade he was dead. He was still in his mid-thirties, struck down by the cholera epidemic that took almost twenty thousand Parisian lives in 1832. To control the contagion, Carnot's body was burned, along with most of his belongings, including a number of unpublished works. His genius would only be realized decades later and the content of his burnt manuscripts will never be known. It is a tale of tragedy that, as we shall see, will be repeated many times throughout the history of thermodynamics.

The story of Julius von Mayer is one such tale. Von Mayer was a physician who served as ship's doctor on an expedition to the Dutch East Indies in 1840. When a sailor took ill, von Mayer would practise bloodletting, opening up the patient's vein to try and relieve the symptoms. It was common practice at the time, but it also led von Mayer to a startling discovery. He noticed that the blood running through the veins of sailors was just as bright red as that running through their arteries. In colder climes, like his native Germany, venous blood travelling towards the lungs would be much darker and velvet in colour. This is because it was lacking in oxygen, which had been used by the body to keep warm through the slow combustion of food. Von Mayer realized that the sailors needed to burn less fuel to stay warm in the tropical sunshine so their veins carried blood with higher levels of oxygen than one might have expected. This meant an equivalence between the heat generated by food in the body and the heat from the Sun. In fact, von Mayer reasoned that all heat was equivalent to *energy*.

With a little bloodletting, the ship's doctor had established the *first law of thermodynamics*: energy is never created or destroyed. It is an eternal shapeshifter, always there, moving from one form to another. He had also identified heat as just another form of energy, in contrast to the old model of caloric that inspired Carnot. Von Mayer wrote up his findings, but his work failed to receive any recognition. Due to his lack of training in physics, his papers were poorly written and littered with errors. The English physicist James Joule independently arrived at the same conclusions and his greater scientific rigour meant that he would take nearly all the credit for the discovery. Von Mayer would suffer again soon afterwards, losing both of his children in quick succession. He fell into depression, attempting suicide and ending his days in mental institutions, a brilliant man broken by personal tragedy and professional neglect.

Nothing can escape the curse of thermodynamics. Eventually, it will touch each and every one of us – every part of the universe in which we live. To understand the terror that is to come, I suggest you make yourself a hot cup of tea. When it is freshly brewed you will notice there is a temperature difference between the tea and the surrounding air. According to Carnot, you should be able to set up a tiny heat engine between the tea and the air to convert the heat into useful mechanical work. You might even be able to drive a tiny little motor. Of course, if you got distracted and left the tea for too long before setting things up, heat would pass from the tea to the air until they both ended up at the same temperature. Then you would be completely stumped – whatever thermal energy there had been available to begin with would suddenly be useless and unobtainable. To get the motor going again, you would need to restore the temperature gradient, but you can't just flip a switch and expect it to appear spontaneously. There is always an energy cost to creating a new temperature difference, and that energy has to come from elsewhere. The easiest thing to do is to boil the kettle and make another cup of tea, but that doesn't come for free.

Something is taking our energy away from us. Of course, this energy is not destroyed, but it is taken out of reach. Who or what is taking it? When a cup of tea is left for too long, what is it that forces the heat to move of its own accord? What is so determined to eliminate the temperature difference, to stop us from extracting energy we can use?

The answer is the entropic captor.

It was the German physicist and mathematician Rudolf Clausius who worked this out after revisiting Carnot's work in the light of Joule and von Mayer's discoveries. Entropy is the agent of heat transfer, the means by which energy is locked away. Clausius billed it as the *transformation content*. That is what entropy means; derived from the ancient Greek *tropos*, which describes a transformation or turning point, particularly in battle. With some clever mathematics, Clausius came up with a formula relating entropy to the energy it imprisons. He found that changes in entropy grow alongside changes to the energy. Furthermore, the entropy was most sensitive to this at low temperatures, when the system was cold.[3]

To see Clausius's formula in action, imagine a kettle powered by a thermonuclear explosion and a brand of tea that can withstand unfeasibly high temperatures. The thermonuclear kettle heats the tea up beyond the temperature of the solar core, to around 100 million degrees Celsius. What happens when a millionth of a billionth of a joule* of heat flows from the tea into the surrounding air? Since it loses some of its heat energy, according to Clausius's formula, the entropy of the tea will fall slightly, by just under a single unit. As the air absorbs the wasted energy, its entropy will rise. The question is: does the entropy of the air rise by more or less than the single unit lost by the tea? The answer is quite emphatic. The air should be around a million times colder than the tea (or you are in serious trouble). This makes its entropy a million times more susceptible to changes in energy – it will increase by almost a million units. The rising entropy of the air completely overwhelms the falling entropy of the tea. The entropy of the combined system – tea and air together – is guaranteed to increase.

This ascent of entropy is known as the second law of thermodynamics. It tells us that the total entropy of a system can never decrease. Sometimes it can stay the same, but in the rusty, turbulent reality of the physical world, it tends to rise, just as it did for the superheated cup of tea. This increasing entropy is the reason that windmills and car engines always lose a little something to their surroundings. The second law can even be applied to the universe as a whole, providing us with an arrow of time, its relentless ascent pointing away from the past and into the future. It is this ascent – this arrow to the future – that I see when I spy a grey hair in the mirror. And it terrifies me, not because of my own ascent towards old age but because of what it all means for the universe. You see, as the entropy of the universe grows, it converts more and more energy into the useless transfer of heat.

* Joules are the standard everyday units of energy. You may be more familiar with kilocalories – one kilocalorie is equivalent to 4,184 joules.

It is smothering our resources, bit by bit, taking away our capacity to extract work, locking away more and more useful energy, like a strait-jacket getting tighter and tighter. The future is a post-entropic nightmare seized by paralysis. This is our heat death, a universe locked in, unable to move, unable to do.

Although Clausius had explained what entropy *did*, he did not explain what it *was*. So, what *is it*? And what could it possibly have to do with doppelgängers? To really understand entropy, we need to peer more deeply into the engines of the Industrial Revolution – we need to look upon the gas within.

The gas is mostly nothing, a vast expanse of emptiness peppered by atoms and molecules, whizzing around without direction. You can imagine them as a swarm of angry insects locked in an empty barn, flying from wall to wall, colliding, falling then rising, from left to right and right to left in random abandon. To picture the gas getting hotter and hotter, you should imagine the flies going faster and faster. Temperature is understood as the average kinetic energy – the energy given to each molecule; here, each insect – on account of its motion. At times the insects collide, bouncing off one another, an elastic encounter on their haphazard journey. They bounce into walls and objects, arbitrarily and erratically but with a collective force that is felt as pressure. If you were standing in the barn, they would collide with you; you would feel their collective touch. If we allowed more insects to join you, they would collide more often, their touch would increase and the pressure would grow. As we filled the barn more and more, that pressure would overwhelm you and destroy you. This is a terror that is known to exist on the planet Venus, where the air pressure is ninety times stronger than on Earth. If you found yourself there, the Venusian air molecules would crush you to death in an instant.

This insectile model of a gas was proposed in 1738 by Daniel Bernoulli, a Swiss prince born into an aristocratic house of science and mathematics that included his father, Johann, and uncle Jacob, early pioneers in calculus and probability theory. Bernoulli's model enabled him to derive Boyle's law governing the relationship between the pressure and volume of a gas, from the mechanics of molecular collisions. Despite this success and his noble position in the scientific establishment, Bernoulli's model was not especially well received. In the eighteenth century most scientists were still invested in the caloric model of heat, with temperature identified as the density of the caloric fluid. They saw little reason for Bernoulli to meddle with heat as a form of energy locked away in the microscopic motion of tiny particles. After all, this was a full century before von Mayer and his bloodletting epiphanies. Bernoulli was simply too far ahead of his time.

To make matters more difficult for Bernoulli, his father, Johann, would then try to steal the work from him, backdating his own later manuscript to look as if it had come earlier than his son's. The relationship between father and son had already been broken by Johann's intense competitiveness. In 1733, the two had shared the Grand Prize of the Paris Academy for individual pieces of work. Johann had been so enraged at the compromise that he cut himself off from his son.

Once caloric theory had died at Clausius's hand, Daniel Bernoulli's brilliance awaited its resurrection. Three men in particular would bear witness: Maxwell, the maestro of electricity and magnetism; the quiet American Josiah Willard Gibbs; and most of all, Ludwig Eduard Boltzmann, a tormented genius who would eventually take his own life.

Clausius, Maxwell, Boltzmann, Gibbs and others began to apply statistical methods to Bernoulli's model. After all, this was a gas with a myriad of particles in random motion, bouncing and bruising their way through empty space. These men showed how collective phenomena could emerge from microscopic chaos. For a gas, temperature and pressure are like the elegant shadows of a murmuration of starlings, absent in the underlying microworld but emerging macroscopically through the power of large numbers. For its part, temperature can be understood in terms of the average kinetic energy of the molecules and how that energy changes with entropy – but what about the entropy itself? What is that?

Entropy is what counts.

I mean this literally. As Boltzmann explained, entropy is really the counting of *microstates*. A microstate is like the ultimate census for a macroscopic object – it tells you everything there is to know about the arrangement of all the atoms and molecules, where they are and what they are doing. When we consider a volume of gas, or even an egg, or a dinosaur, we know it is made up of lots of little things. Each atom is here or there, spinning this way or that, moving at some particular speed across short bursts of space, and there are billions upon billions of them. The atoms themselves have their own building blocks, of course, with their own intrinsic properties. To fully describe the gas, the egg or the dinosaur you could – if you were crazy – write down a gigantic data array, listing the position, velocity, spin, favourite colour, favourite box set and whatever else, for every single one of the billions of building blocks in the system. Such an array would describe a particular microstate, giving you complete and precise information about the object in question.

But here is the thing: if you change the positions of a few atoms here and there, no one is going to notice. The egg will still look like exactly the same egg, the volume of gas will still have the same temperature, the dinosaur will still be a triceratops that should have died 65 million years ago. The point is that when we look at big things it would be silly to worry about all the detailed minutiae. Entropy is a measure of this hidden detail. It counts up all the microstates that leave the macroscopic properties of an object unchanged. As time goes by and the egg or the dinosaur begins to crumble, as it disintegrates into dust, more and more of its microscopic details are hidden away. Staring at the dusty remains, it becomes increasingly difficult to distinguish between one possible microstate and another. With a troubling inevitability, the microstate count of the egg or the dinosaur increases over time. This is the ascent of entropy, a count that goes up but never comes down.

Entropy doesn't always have to be about molecules and atoms. We can talk about entropy in any context as long as there are microstates of some sort and they can be counted. Take facial recognition software, for example. Thankfully, my phone recognizes me as me even though I don't always adopt the exact same facial expression as when I first logged in. It does away with all the superfluous data and identifies lots of subtly different images of me as being one and the same thing. If you were to count them all up, that would be a measure of the entropy of my face.

Here is a more quantitative example: in the English Premier League there are twenty football teams, playing each other home and away over the course of a season. That's a total of $20 \times 19 = 380$ games per season with each game having one of three possible outcomes – home win, away win or draw. This means that there are 3^{380} ways in which the season can play out in terms of results. However, many of those 3^{380} outcomes will yield precisely the same league table in terms of the number of points attained by the champions, the runners-up, and so on. We can think of the various outcomes as microstates and, for a given end-of-season table, we can count up all the ways in which we could have ended up with the same distribution of points across the league. This would give us a measure of Premier League entropy.

With twenty teams, the Premier League mathematics is far too painful to look at in any further detail, so let's reduce the number and imagine a league trimmed down to its two biggest rivals: Liverpool and Manchester United. The rest of the best have all been relegated in the interests of mathematical simplicity, including Everton, Arsenal, Spurs and even the oil-rich football state of Manchester City. In this shrunken Premiership, there are only two games to play over the course of a season and therefore a total of nine different outcomes. If we decide not to worry about who comes first and who comes second, then different outcomes can yield the same league table. Remembering that there are three points for a win, one for a draw and zero for a defeat, the nine possible outcomes spread across four distinct tables, as shown in the figure below.

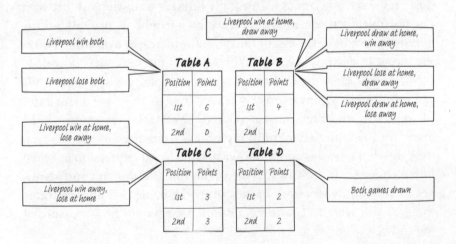

Let's take a closer look at table A, for which the champion scores six points and the runner-up scores none. This can be realized in one of two possible ways: Liverpool win both games, or they lose both games. In other words, there are two distinct microstates that yield the same league table. This counting gives a measure of the *entropy* of table A or, more precisely, its natural logarithm does.

I should quickly explain what a logarithm is. A logarithm of a number brings down its power with respect to some particular base. For example, if we choose base 10, then the log of 100 is 2, since 100 is 10 to the power of 2. For a *natural* logarithm, usually written as 'ln', the chosen base is *Euler's number*, $e \approx 2.718$, and so the natural log brings down the power of e. For example, $\ln e^2 = 2$, $\ln e^3 = 3$, $\ln e^{0.12} = 0.12$, and so on. Natural logs are much more prevalent in science than base 10 logs.

Boltzmann proposed a formula for entropy given in terms of natural logs, $S = \ln W$, where W counts the corresponding microstates, or the number of *ways*. Returning to the shrunken Premier League, it follows that the entropy of tables A and C both go as $\ln 2 \approx 0.693$; table B as $\ln 4 \approx 1.386$; and the entropy of table D is zero (since $\ln 1 = 0$). We count states and entropy in exactly the same way when we are talking about eggs or dinosaurs. The only difference is in the numbers involved. The number of microstates that could describe the egg (or dinosaur!) you had for breakfast is vast, up there with the googols, as opposed to the one, twos and four we had for our two-team Premier League outcomes.

Now that we have a notion of Premier League entropy, how do we see its likely ascent? It's actually quite easy. Imagine that the season finished with table A and an entropy of $\ln 2$. What will happen the following season? If all individual outcomes are equally likely, then there is a nine-to-four chance that the entropy will remain at $\ln 2$ (tables A and C combined), a nine-to-four chance that it will increase to $\ln 4$ (table B) and only a nine-to-one chance that it will drop to zero (table D). Thus, even in this small-scale example, the entropy is much more likely to go up than go down.

When we scale the numbers up to googolist proportions for the atoms in an egg or a dinosaur, the dominant probabilities become overwhelming. The ascent of entropy is no longer likely but *inevitable*. Imagine a cube of ice at room temperature. This system is described by an icy microstate or microstates and, over time, it will move to other possible microstates. The system makes several microscopic hops between states and no one is surprised to find that it ends up a puddle. There was a very slim chance it could have remained an ice cube, but it wasn't at all likely. At room temperature, the paucity of icy microstates in comparison to puddly microstates just meant it was overwhelmingly likely that the ice would melt. Entropy's relentless ascent is really just the inevitable rise of the masses.

By playing these statistical games, we can also understand the laws of thermodynamics, where energy is taken prisoner by its entropic captor and the universe is headed towards paralysis. The point is that the more microstates you accumulate, the more your knowledge of an egg or a dinosaur or a puddle is diluted. In a sense, it becomes more difficult to steal the useful energy because you can't quite be sure where it is held. It's a bit like a burglar trying to steal a precious jewel: if it is held in a large mansion with hundreds of rooms, chances are it will take them a long time to find it. If the mansion is big enough and the burglar is chaotic in their approach, they may never find it. So it is with entropy, hiding the energy in a blur of confusion, making it harder and harder for us to steal. Boltzmann understood that if you left things alone to manage themselves, confusion and ignorance would always increase. Spend some time watching the news and listening to our politicians and you will quickly realize that Boltzmann was right.

Boltzmann's work was truly remarkable. He didn't just leap recklessly from the microscopic to the macroscopic, from Lilliput to the giants of Brobdingnag. He built a bridge with strong mathematical foundations and showed exactly how one should walk safely across. Of course, as ever, his ideas met with some resistance because not everyone was ready to accept the reality of atoms and the dominance of empty space. Boltzmann was not especially well equipped to deal with this resistance. For all his brilliance, he was a deeply troubled man, prone to violent mood swings, manic behaviour and deep depression. It ended with another thermodynamic tragedy, at Duino, near the city of Trieste, when Boltzmann hanged himself as his wife and daughter swam in the bay. He left no note. It is not known if his professional difficulties led him towards this desperate act. What we do know is that a year earlier, unbeknownst to Boltzmann, Einstein had published work that would ultimately convince the scientific world of the reality of atoms and to follow Boltzmann's bridge into the macroworld.[4]

Let's return to you and your doppelgänger. Like the egg, the dinosaur and the volume of gas, you are also made up of billions and billions of atoms and molecules. It's impossible to know exactly where all the atoms are and what they are doing. As a result, there isn't just one arrangement, one data array, that could do a perfectly adequate job of describing you reading this book in your macroscopic world. There are *a lot*. Of course, there are also all those other microstates that have nothing to do with you reading this book. There are those that describe you reading a copy of *Hello!* magazine, those that describe a cow reading *Hello!*, those that describe a gas of molecules at a given temperature and pressure, and even those that just describe empty space. In fact, for the cubic metre of space you more or less occupy, we could imagine an infinite number of different scenarios – subtly different versions of you, cows, gases, vacua. So, there must be an infinite number of microstates that could, in principle, describe any given cubic metre of space, right?

Wrong.

This number is *finite*. If it were infinite, there would be nothing to stop the entropy in this cubic metre from growing and growing, from a googol to a googolplex, to TREE(3) and beyond. But something does stop it: gravity. Clausius taught us that entropy and energy grow side by side, and Einstein taught us that energy weighs. If you try to squeeze too much entropy into a cubic metre of space, gravity will feel the weight of the corresponding energy, and summon the gaoler. A black hole will inevitably form.

Black holes are the entropic limit. They disguise their microscopic secrets better than anyone or anything. They are the faceless passer-by whose terrible history you will never know – *can* never know. When you look at them, when you try to measure them, black holes will only ever reveal three things of themselves: their mass, their charge and their spin. Everything else will stay hidden. Imagine you encountered a small black hole at the bottom of your garden; would you know how it came to be? If it were still there a day later but had grown more massive, by about the mass of an elephant, could you be sure it had consumed an elephant? Could it not have consumed the complete works of Shakespeare with elephantine mass, charge and angular momentum? Both scenarios would result in the same black hole, with the same three characteristics, so how could you possibly know which of the two had actually occurred? How could you know the true history of the black hole?

This black hole secrecy hints at its unrivalled ability to store entropy. There are many ways in which it came to be, be it through elephants or Shakespearian texts, and yet none of this is encoded in its macroscopic features. It is lost in the community of possible microstates, whatever they may be. For a given volume of space, nothing is more entropic than the black hole that just fits inside this volume, with its event horizon touching the outer edge. But if black holes are the entropic limit, how much entropy do they have?

Well, for most macroscopic objects like eggs or humans or dinosaurs, entropy grows with volume. For example, a mother triceratops that is ten times larger than its baby in all directions will have about a thousand times more entropy. This makes intuitive sense: the mother occupies a volume one thousand times larger and therefore has space for up to a thousand times as many atoms. Each atom allows for a handful of new possibilities. For example, the atom will be spinning this way or that way. That is two possibilities for each new atom. For a hundred new atoms we would have 2^{100} possibilities; for a million new atoms there are $2^{1000000}$ possibilities; and so on. Clearly the number of possibilities, the number of microstates, grows *exponentially* with the number of atoms. The entropy is the logarithm of this – it brings down the power – so that must be *proportional to* the number of atoms. That is a thousand times greater for the mother triceratops compared to her baby.

But a triceratops is not an especially entropic object. We could squeeze a billion triceratops into the same space and create a triceratops-crush of far greater entropy but with the same volume. Eggs, humans, triceratops – none of them are anywhere near the top of the entropic food chain. But black holes are, and it just so happens that the entropy of mother and baby black holes scale very differently to that of our mother and baby triceratops. Black hole entropy grows with the *area* of the event horizon, rather than the volume. This is completely counterintuitive but only because we aren't used to dealing with objects that are so overwhelmed by gravity's crushing embrace.

In the early 1970s, the Israeli-American physicist Jacob Bekenstein and his British counterpart, Stephen Hawking, showed that a black hole with an event horizon of area A_H has entropy given by

$$S = \frac{A_H}{4l_p^2}$$

The symbol l_p represents the Planck length.* It is the shortest meaningful length in physics – around a billionth of a trillionth of a trillionth of a centimetre. It corresponds to the point where we start to lose control of our understanding of gravity – when gravity starts to flirt with the microworld of quantum mechanics, where the fabric of space and time becomes hazy, blurred and maybe even broken.

Hawking was able to pin down the details of this formula using some clever thermodynamic arguments, but a proper *microscopic* derivation is still lacking. What we'd really like to do is take a typical black hole and identify all the microstates consistent with its three macroscopic properties – its mass, charge and spin. We'd then count the microstates and see if the resulting entropy matched Bekenstein and Hawking's formula *exactly*. No one has figured out how to do this yet, at least not for the kind of black holes you find patrolling the centre of a galaxy.[5] It remains the holy grail of black hole research.

Let's return to the cubic metre of space that you happen to occupy, or indeed *any* cubic metre of space. How many microstates do you need to be absolutely certain of capturing all its physics? To answer this, we need to touch upon all possible microstates and push things to the entropic limit. In other words, we need to consider the largest black hole that could fit inside the space. Such a black hole would have an event horizon with a surface area of around a square metre and therefore, according to Bekenstein and Hawking's formula,[6] an entropy of around 10^{69}. That equates to around $10^{10^{68}}$ microstates. This is it. This is the limit. This is the largest number of microstates you will ever need to describe a cubic metre of space.

As an aspiring googologist†, I'm going to give this enormous, but finite, number a name: the doppelgängion. We have found it as we bridge between this chapter and the next, between a googol and a googolplex. It feels appropriate. After all, the doppelgängion lies somewhere in between these two leviathans. It towers above a googol but is well short of a googolplex. To fully appreciate its significance, we need to continue the search for your doppelgänger deep into the

---

* The precise value of the Planck length is 1.6×10^{-35} metres.

† A googologist is someone who studies and invents names for large numbers.

next chapter, exploring what it means to be you all the way down to the subatomic realm.

Thanks to this entropic limit, I now know that the cubic metre I occupy as I write these words is described by at least one of up to $10^{10^{68}}$ possible microstates. This is also true of the cubic metre of space occupied by Prince Harry and Meghan Markle, or by a gaseous alien plotting intergalactic war off the coast of Andromeda. And it's true of you. You're better than one in a googol, but you're not one in a googolplex. The best any of us could be is one in a doppelgängion.

Perhaps I'm being too kind. From a choice of $10^{10^{68}}$ microstates, there will be several that could adequately describe you and your macroscopic features – same nose, same ears, same delighted expression, and so on. Presumably your doppelgänger samples the same states. If we wanted to be more precise, we could try to narrow down the relevant microstates a little further. We could start asking about the precise states of individual atoms in your body, or of the neurons firing thoughts around your brain. It all depends on how carefully we want to define you and, by association, your doppelgänger. How exact does the match have to be for us to call it? Is it enough for you to look alike, or do we also require you to have the exact same thoughts and the same arrangement of atoms? However, the moment you start measuring the states of individual atoms, you are entering the microworld, the realm of quantum mechanics and the subject of the next chapter. The quest to find your doppelgänger has just become a quantum quest. To be honest, it was always so – the universe is quantum. *You* are quantum.

And so is your doppelgänger.

A Googolplex

THE QUANTUM SORCERER

You've had a few too many, but it doesn't matter. Wednesday night is quiz night down at your local pub and tonight there was a question about entropy. You were the only one that knew the answer, so you are feeling rather pleased with yourself. As you stumble home, you sense a passer-by on the other side of the road. Wait. He could be on the same side. Or is he in the middle of the road? You can't tell. What on earth is going on? Did you really drink that much?

Welcome to the microworld, where every passer-by is a sorcerer, where quantum mechanics is lord, and where you are here, there, everywhere and nowhere, lost in a fog of probability. You may be surprised that I should bring you here, to the smallest of worlds, when our ultimate goal is to imagine a googolplex – a one followed by a googol zeroes – and the vastness of a googolplician universe. But I have no choice. If you are to properly appreciate the googolplician universe – to identify the doppelgänger that lies within – you will need to understand the quantum laws. They are like nothing you are used to. They are strange and counterintuitive. But to continue on our journey, we will need to learn a new way of life. It is a life that exists beneath our everyday existence in the dance of the subatomic particles that make up each and every one of us – a dance that makes you, you and your doppelgänger, you.

Quantum mechanics grew from the rubble of catastrophe. By the end of the late nineteenth century physicists were openly triumphant. They had ushered in an era of discovery and invention: electricity, magnetism, light, radio, atoms, molecules and thermodynamics. Their genius lit up the streets of London, Paris and New York, it drove the engines of the Industrial Revolution and it was on its way to changing the world with radio and television. But all was not well. There was a fly in the ointment, an embarrassing secret, an absurdity born from their best and most trusted ideas.

The ultraviolet catastrophe.

When a physicist talks about the ultraviolet, they just mean something that is oscillating with a really high frequency. For example, you've probably heard of ultraviolet light – that's just like visible light, but with a frequency that is too high for us to see. The ultraviolet catastrophe emerged when nineteenth-century physicists began to think about how much energy would be stored in the high-frequency radiation absorbed or emitted by certain objects. You can experience this catastrophe in the comfort of your own home.[1] Suppose you have an oven in the kitchen that is perfectly insulated and you turn the dial up to 180 degrees Celsius. By the time it's up to the right temperature, you can ask: how much energy is stored in the oven? To figure this out, you take a look inside the oven. It looks empty, but you know it's not really empty. It is filled with waves of electromagnetic radiation, wriggling like Maxwell's sea snakes from chapter '1.0000000000000000858'. You notice that some snakes wriggle more furiously than others, completing more oscillations from head to tip. There is energy in these oscillations and you begin to add it all up. With a little help from the ghost of a late-Victorian physicist, you are able to calculate the total energy in all the oscillations.

The answer you get is *infinity*.

No wonder the Victorian ghost looks embarrassed. He should be. It's a catastrophic answer. How did he screw up so badly? To investigate what has happened, let's look at an individual wave of electromagnetic radiation. We can think of this as a twin pair of sea snakes – the electric snake and the magnetic snake – trapped in the oven but wriggling to and fro at right angles to each other, as in the figure on the next page.

This wave has two important features: the *frequency* of the oscillations and the *amplitude*. The frequency tells us how fast the snakes are wriggling, and the amplitude is the height of their wriggles. The Victorian ghost paints you a picture of a great many snake pairs, all

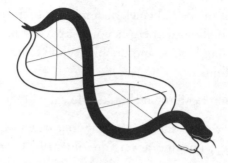

wriggling with the same amplitude but with a full range of frequencies. He also tells you what Maxwell and Boltzmann told him: that, on average, the energy stored in each snake pair is the same – *it does not depend on the frequency*. In fact, he convinces you that each pair carries around 6 *zepto*joules* of energy,[2] which is 100 trillion trillion times less than the 200 calories you would get in a Mars Bar. Despite this tiny amount, he warns you that the full range of frequencies is in fact infinite. That must mean an infinity of wriggling snakes, filling the oven with an infinite amount of energy. His logic has led you straight into the ultraviolet catastrophe and an infinitely large energy bill.

But there is no need to panic just yet. The truth is that we now know how to avoid the catastrophe, thanks to the brilliant German physicist Max Planck. Like many of the protagonists in this book, Planck would suffer in his personal life – his son, Erwin, was executed by the Nazis for his part in Claus von Stauffenberg's failed assassination attempt on Adolf Hitler.

Planck realized that not all snakes were born equal – the energy they contributed had to depend on how fast they were wriggling. If he were to avoid the ultraviolet catastrophe, the wriggliest snakes, on average, had to contribute less and less energy to counter the fact that there were infinitely many of them. Planck figured out how this could actually happen. Electromagnetic waves could no longer have any amount of energy (as our Victorian ghost had assumed). There needed to be gaps in the energy spectrum, growing larger and larger as the frequency increased, suppressing the average. To match the experimental

* A *zepto*joule is a billionth of a trillionth of a joule, or 10^{-21} joules.

measurements of the day[3], Planck also noticed that the gaps had to be very precise. The allowed energies could appear only in very specific chunks, or building blocks, and the higher the frequency of the wave, the bigger the chunks.

But Planck didn't call them chunks. He called them *quanta*.*

To get a better understanding of the mathematics behind Planck's chunky epiphany, imagine a version of *Squid Game*, where debt-ridden contestants risk their lives playing children's games in the hope of winning an enormous cash prize. Suppose there are 511 players with varying levels of debt:

- 1 player owes 8 billion Korean won (KRW)
- 2 players owe 7 billion KRW
- 4 players owe 6 billion KRW
- 8 players owe 5 billion KRW
- 16 players owe 4 billion KRW
- 32 players owe 3 billion KRW
- 64 players owe 2 billion KRW
- 128 players owe 1 billion KRW
- 256 players are debt-free

At the start of the contest the average debt of the players is just short of a billion KRW (982,387,476 KRW, to be precise). By the end of the first game, all those owing 1 billion KRW, 3 billion KRW, 5 billion KRW or 7 billion KRW have been brutally 'eliminated'. There are fewer players now, but their total debt is significantly reduced – the average debt of the remaining players falls to around 657 million KRW. By the end of the second game, those owing 2 billion KRW and 6 billion KRW are also eliminated. The average debt of the remaining players is now just 264 million KRW. After each new game, players are eliminated and larger gaps appear in the 'spectrum' of debt, bringing down the average.

Planck realized that something similar must be happening with the waves in your oven. When you perform an energy census for waves of a certain frequency you find that the oscillations take up energies only

---

* Quanta is the plural of *quantum*, Latin for 'how much' or 'how many'.

in chunks of a particular size. For higher frequencies, the chunks get larger and the average energy drops like a stone.

To match with experimental data, Planck calculated that waves of frequency ω must have energies that are integer multiples of $\hbar\omega$, where $\hbar$ is a super-tiny number – the so-called Planck constant – less than a billionth of a trillionth of a trillionth in everyday units.* As we will see in a moment, the smallness of $\hbar$ is the reason the quantum world stayed hidden from us for so long.

In a way it is strange that waves are straitjacketed in this way, forced by natural law to pick a very particular set of energies, depending on their frequency. For example, according to these rules, waves of frequency 10^{33} Hertz are allowed to pick their energy up only in integer bundles of joules: 1 joule, 2 joules, 3 joules, and so on. All other energies are completely forbidden, which begs the question: what happens when I try to feed one of these waves half a joule of energy? Wouldn't that take it out of the allowed range and spark a revolution? Indeed it would, and for that reason the wave would refuse the meal! Its respect for the law is absolute, and the underlying chunks, or quanta, will always remain sacrosanct.

These chunks of $\hbar\omega$ use the currency of the Planck constant much in the same way as the Korean won is used as monetary currency in *Squid Game*. Because the Planck constant is super-tiny (in everyday units), it took us an awfully long time to notice that the chunks were there in the first place. It is the same with money – you wouldn't notice the difference of a single won if all you ever traded in were billions of won. Initially, Planck saw these chunks and his currency as little more than a mathematical curiosity. However, the reality was that his mathematical incantations had torn open a portal, revealing profound truths about the physical world, just as Maxwell had done half a century earlier when studying the mathematics of electricity and magnetism. That said, it took the courage of Albert Einstein to climb through and tell the world what Planck had exposed.

To properly appreciate this, we need to talk about a little experiment where you shine a beam of ultraviolet light on a zinc plate and

---

* The precise value of the Planck constant is given by $\hbar = 1.05 \times 10^{-34}$ joule seconds.

the metal begins to spit out electrons. That's not particularly unusual. Ultraviolet light can do terrible things to you, as I can testify whenever I forget to apply sunscreen. What is strange about this experiment is what happens when you increase the *intensity* of the light. You might expect the electrons to be spat out with greater speed because there is more energy in the beam. But that doesn't happen – you get more electrons, for sure, but the speed with which they are spat out remains the same. The only way to get faster electrons is to increase the *frequency* of the beam. X-rays have a higher frequency than ultraviolet rays. This means that an X-ray beam will produce faster electrons than an ultraviolet beam, even if the X-rays are less intense. The reverse is also true: if you drop the frequency of the beam, the electrons slow down, and if you drop it enough, they stop being produced altogether. Shining visible light on zinc will not produce any electrons because the frequency is too low.

Einstein had an explanation for this idiosyncratic set of results, commonly known as the *photoelectric effect*. The year was 1905, his *annus mirabilis*. Although he would propose special relativity in the same year, he always saw his work on the photoelectric effect as more revolutionary, more rebellious against the established lore. We can understand the spirit of his rebellion with another analogy, albeit a drunken one. Picture yourself in a crowded vodka bar, filled with a googol thirsty patrons waiting to be served. Right now, they are sober, but after half a litre of vodka they can expect to be classed as drunk. As soon as that happens the bouncers will throw them out on to the street, where Einstein is watching events unfold. A delivery of vodka arrives, and it takes the form of a few thousand 50-millilitre miniature bottles. These customers are a selfish bunch, and so there won't be any sharing. The bartenders distribute the miniatures randomly but, because there are so many customers, most end up with nothing. Some get hold of a single bottle, but it's very unlikely that anyone will be lucky enough to get any more than that. As a result, there won't be any individuals with enough vodka to get a bit tipsy and nobody will be thrown out. The next day, a delivery arrives with a billion 50-millilitre bottles, but it makes no difference – no one gets their hands on enough vodka to get drunk enough to be thrown out. On the third day, the vodka company decides to up the ante. They do away with the miniatures and deliver litre bottles instead.

There are a few thousand of them and, once again, they are distributed randomly by the bartenders. After a while, Einstein finally starts to see people being thrown out. They are visibly drunk and without exception they are each carrying a litre bottle of vodka that is half full. On the fourth day, there is another delivery of litre bottles, only now there are a million of them. Einstein sees many more drunken revellers cast on to the street but, again, each of them is holding a bottle of vodka that is exactly half full.

What does all this hedonism have to do with the photoelectric effect? Einstein realized that if light were split into chunks, as Planck had suggested, then the photoelectric effect could be easily explained with something akin to our hedonistic analogy. You can think of the bar as being like the metallic plate, the customers as the electrons and the vodka delivery as the beam of ultraviolet light. If Planck were correct, light would be forced to deliver energy in chunks of a definite size depending on their frequency, just as the vodka was always delivered in 50-millilitre bottles or litre bottles. Whenever a single chunk was delivered to a zinc plate, 700 zeptojoules would be needed to dislodge an electron and anything left over used to speed it up. As the size of the chunks was always fixed, the energy left over was always the same and the electrons were always sped up by the same amount. It didn't make any difference if you increased the intensity of the beam – that just delivered more chunks, so more electrons would be dislodged but all with the same speed as before. It was the same with the vodka. When the litre bottles were delivered, it didn't really make a difference how many. All that mattered was that it was enough to cross the half-litre threshold of drunkenness, and anyone who hit this threshold was guaranteed to be thrown out with half a litre left over. You can also see why the electrons stay put when visible light is shone on a zinc plate. Take blue light as an example – it will be delivered in chunks whose size is around 400 zeptojoules, which is simply not enough to dislodge an electron.

The photoelectric effect proved the chunkiness of light. The chunks, or quanta, of light are known as *photons*, legislated to carry a very definite amount of energy, like a single worker ant tasked with carrying a very specific leaf of a very specific size. This was incredibly unsettling. For over a hundred years, since the pioneering experiments of the British polymath Thomas Young, light had been firmly established as a wave, and yet here it was, behaving like a particle. This would be like waking up one morning to hear that Greta Thunberg had endorsed Donald Trump. You just wouldn't see it coming.

Young had established the wave-like properties of light in a classic experiment, cutting two slits very close together in a dark screen and shining through a beam of light. He placed a second screen behind the first and detected the image left by the light that had passed through. If the beam were a spray of particles, Young had expected to detect an unbroken band of light, the centre of maximum intensity lying in the middle region directly behind the two slits. You can think of this like a hail of bullets fired indiscriminately towards the screen. They will be deflected as they pass through the narrow gaps and more will gather in the mid-zone than out to the sides. The mid-zone is the worst place to stand because bullets are received from both directions, in contrast to the far right, say, where you would be in danger only from bullets that passed through the right-hand slit. But this pattern of bullets is *not* what Young saw in his experiment with light. Instead, he detected a series of light and dark bands that resembled a supermarket bar code.

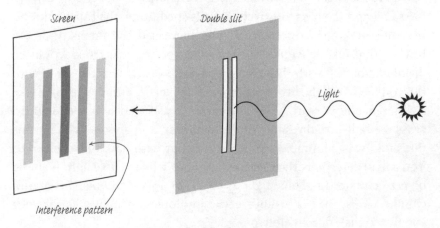

Young's double-slit experiment.

This was consistent with the light passing through both slits at once like a tidal wave breaking through adjacent doors on a beachfront hotel and then interfering with *itself* on the other side. The dark bands could be understood as peaks and troughs of different waves interfering destructively, pushing in opposite directions to cancel each other out. In contrast, the light bands coincided with the waves interfering constructively, pushing in the same direction, their collective effort accumulating into much brighter regions. The pattern of bands was unmistakable: Young's experiment pointed to light behaving very much like a wave rather than a particle. And now the photoelectric effect seemed to suggest the opposite.

So, which is it? A wave or a particle?

The truth is that light is like the ultimate stage actor. It can change its outfit depending on the show. When Thomas Young is producing and the stage is set with a double-slit experiment, light will dance like a wave. When the show is being produced by the photoelectric company, light will dance like a particle.

Now you might think you can explain this away by saying that photons are particles and that the wave-like behaviour of light is nothing more than a macroscopic effect – a property of their murmuration. After all, water waves are really made up of lots of tiny water molecules, so perhaps when you have a large enough colony of photons they conspire to behave more like a wave. Actually, this colony of photons is a very good way to think about an everyday beam of light. But here is the thing: Young's experiment still yields the same results even when you drop the intensity of the beam to an inexplicably low level, firing one photon at a time. What happens is that the individual photons each land on the screen at a random location but, eventually, the bar-code pattern begins to emerge. When the stage is set with the double-slit experiment even a single photon will dance like a wave. This is one of my favourite facts in all of physics – an individual particle of light acting like a wave, almost as if *it passes through both slits at the same time*. It is absolutely mind-blowing and has no right to be true. But it is!

There is just no escaping this fact: a single photon can behave like both a particle and a wave, depending on the mood. But what about things we normally think of as particles, like electrons and protons? Could they not be waves too? Of course they can. Light is not the only actor on stage – it turns out that matter is more than capable of putting on exactly the same sort of show. When two American physicists,

Clinton Davisson and Lester Germer, fired electrons at a narrow pair of slits, the electrons gathered on the screen behind, painting an image of a supermarket bar code, as every self-respecting wave is guaranteed to do.

By the time Davisson and Germer had completed their experiments in the mid-1920s, the results had already been anticipated. The stage was set more than a decade earlier, by New Zealand's most celebrated physicist, Ernest Rutherford, or to give him his full title: The Right Honourable Lord Rutherford of Nelson, Order of Merit. As his title suggests, Rutherford was an important man, a Nobel Laureate and the father of nuclear physics. In the years before the Great War, Rutherford's experiments established that atoms resembled a miniature solar system, with planetary electrons in orbit around a dense core dubbed the nucleus. The broad cloud of electrons carried a negative charge, in contrast to the nucleus, which was a concentration of positive charge. This charge meant that the dynamics of the atomic solar system were dominated by the electromagnetic force. But to Max Planck, Rutherford's model made no sense – orbiting electrons were accelerating, and by Maxwell's theory that meant that they should radiate away their energy and fall into the nucleus almost instantaneously. The atom should really be nothing more than a tedious clump of neutrality. It had no right to exist.

Over in Copenhagen, the problem had caught the attention of an ex-footballer by the name of Niels Bohr. In his teenage years, Bohr had played in goal for the Danish team Akademisk Boldklub, alongside his brother Harald. While Harald went on to play for the national team at the Olympic Games, Niels decided to concentrate on physics and, by 1913, he had figured out how to save the atom.

Bohr took some of Planck's own currency, his tiny constant $\hbar$, and argued that the electron orbits should be distributed in very precise chunks. Because of this chunkiness, orbits could not be arbitrarily small, and, for the hydrogen atom, he calculated the smallest one to have a radius of around fifty trillionths of a metre. The next level up would be at four times the radius; the next one at nine times; and so on. You can picture Bohr's atom as being like an apartment block where the ninth floor is overrun with zombies. If the zombies reach the ground floor the city will be destroyed. To safeguard against this, the authorities have closed off the stairwell and reprogrammed the lift so that it can only stop at certain floors. It was already on the ninth floor, but they succeeded in making it so that the only other floors it could stop at would be the first and the fourth. After a while, some zombies stumble into the lift and end up elsewhere in the apartment building, occasionally reaching the fourth floor and sometimes as low as the first. But they never get any lower. They never reach ground level because the lift doesn't stop there. The city survives. So it is with the atom. Once an electron has reached the lowest level allowed, calculated using Planck's own currency, it is forbidden from going any lower and the atom is allowed to exist.

Although Bohr had established a set of rules, he didn't really explain why the electron should obey them – why it would orbit the nucleus at such precise locations. Enter a young French prince: Louis de Broglie, 7th Duke of Broglie. In 1924, he submitted a PhD thesis to the University of Paris, arguing that Bohr's atomic orbits could be understood as an electron wave rather than a particle, wrapped into a circle, like an image of the serpent Ouroboros, wriggling its way around to eat its own tail. Depending on the momentum of the electron, the corresponding wave should have a very particular wavelength.[4] The wavelength is just the distance between neighbouring peaks or troughs in the serpent's wriggle – high-momentum particles have short wavelengths, whereas ones with low momentum have long wavelengths. In order for peaks and troughs to line up neatly as you go once around the circle, there must be an integer number of them. That could only happen at a very discrete set of radii. It's a bit like a party of people dancing in rings around one another. They form circles made up of different numbers, each with their arms outstretched, holding hands with their nearest neighbours. The innermost ring is the smallest, comprising just one dancer – the baby of the troupe – holding hands with herself. Next, there are two teenagers: their arms are twice as long as those of the baby and so the circle they form is *four* times larger (remember, their arms are longer and there are two teenagers, each with two arms). The third ring is made up of three adults – their arms are three times longer than the baby's and so their ring is *nine* times larger. If the troupe were to recruit giants with even longer arms, four times the size of the baby, we could go on. The point is that by forcing them to connect in a circle the various members of the troupe find themselves dancing at a very particular radius, just like electrons dancing in an atom.

Despite his modest academic status, the young PhD student de Broglie had caught the attention of Einstein, who immediately recognized the importance of his ideas. De Broglie had sparked a revolution. An army of brilliant young physicists rushed to enlist, ready to challenge the established lore, people like Werner Heisenberg, Erwin Schrödinger, Pascual Jordan and Paul Dirac. The Austrian Schrödinger was one of the first to stand up. He had been inspired by a flippant observation* he'd heard at a conference, that a wave-like electron should satisfy some sort of wave equation. To tackle the problem, he left his wife at home over Christmas and took off to a remote cabin in the Alpine resort of Arosa in Switzerland. He packed a copy of de Broglie's thesis and invited along his mistress from Vienna for company. It was a scandalous few weeks, but by the end of it Schrödinger had discovered one of the most important equations in physics.

Although Schrödinger was able to reproduce the correct physics of the hydrogen atom using his wave equation, it wasn't exactly clear what the wave actually was. Schrödinger christened it the *wave-function* and was convinced that it described a spread in the charge of the electron, as if it were smeared out over space. But this wasn't right. Although Davisson and Germer saw that the electrons built up a wave-like pattern in their version of the double-slit experiment, the pattern emerged only after a great many electrons had hit the screen. The reality was that an *individual* electron always landed at a single random location – its individual charge was never broken up and spread across in a bar-code pattern, as Schrödinger was trying to suggest.

It was Max Born (Olivia Newton-John's Nobel Prize-winning grandfather) who realized what was actually going on: Schrödinger's wave-function is a wave of *probabilities*. Its overall magnitude tells you where the electron *could* be and how likely it is to be there. If you looked for the electron, chances are you would find it near where the magnitude peaks, but there is no guarantee – it could be anywhere that

---

* The observation is usually attributed to the Dutch physicist Peter Debye.

the wave didn't vanish. Until you perform a measurement and pin down the electron, you cannot know where it is. It is left to chance.

It's a bit like trying to keep track of a fugitive using a cheap GPS tracker. You cannot pin down their location exactly. The best you can do is say that they are hiding somewhere in the shopping mall in town, probably in the middle of it, but you cannot know for sure. The true location is left to chance. You can distribute police officers at strategic locations around the mall, but you cannot know which one will actually catch the fugitive. It is only when they are caught that you know for sure where they are. It seems as if nature has condemned us to using cheap GPS trackers. In the double-slit experiment, the final location of an individual electron is left to chance and it is only after you perform lots of measurements and capture lots of electrons that a pattern starts to emerge consistent with the wave of probabilities. The implications are profound.

Determinism is dead.

In other words: the past cannot completely determine the future. We know this is true of the electron in Davisson and Germer's experiment – its fate is fundamentally *unknowable*. It could land here or there with some *probability*, but there is never any way of knowing for sure. God simply loves to play dice. Nature is a game of chance. If you are unlucky in love, do not despair that you are destined to be alone. Just remember that there is no such thing as destiny in the microworld.

Perhaps the most important thing about these waves of probability is the way they can be superimposed on each other. This is true of any wave. If you are on board a ship and you throw a stone over the side, it will land in the water and create ripples. Those ripples superimpose themselves on the large ocean waves that swell up and down against the side of the ship. This superimposition is known in physics as *superposition*. For the double-slit experiment, what you get is the probability wave for an electron passing through the left-hand slit superimposed on the probability wave for an electron passing through the right-hand slit. The end product is a wave that democratically combines the two, trading the ripples of one against those of the other, consistent with the beautiful bar-code pattern detected on the screen.

We now know that when an electron lands on the screen in the double-slit experiment, it lands where it lands with a certain probability. We know where it started and where it finished, but do we know how it got there? Did it pass through the left slit or the right? We cannot know for sure, and that's why we talk about probabilities, but common sense would surely suggest that it passed through one or other of the slits.

Richard Feynman wasn't so sure.

Feynman was a rock-star physicist, blessed with charm, good looks and a razor-sharp New York accent. He was also a genius. In the years that followed the Second World War, he argued that a wave-like electron could be understood as going through *both* slits simultaneously. It's not that it spreads itself out in the way that Schrödinger envisaged – it's much weirder than that. It literally goes one way *and* it goes another.

And another.

And another.

In fact, the electron goes in every which way you could possibly imagine. It doesn't just take the most well-trodden paths through each of the slits. It also takes nonsensical paths, like those that break the cosmic speed limit looping around the furthest point of Andromeda or burrowing towards the centre of the Earth before burrowing back up again. Feynman's perspective says that there is a sense in which the electron does all of these things and more. But here is the really clever bit. He also showed how to assign a particular number to each particular path between the two points. When you work out the average of these numbers over all the different paths, you get the probability wave for an electron going between those two points. There is no need to put the electron wave in by hand – just assume all possible paths or, in other words, all possible *histories*, and sum them up.

This also applies to you when you take a walk to the shop at the top of your road. You may think you walk directly from your house to the shop, but that is only one path. In reality, you explore all possible paths, including those that wander off to each and every corner of the universe. Of course, in your particular case, the overwhelmingly dominant contribution to your 'sum over histories' comes from the deathly dull path direct from your house to the shop. This happens because a macroscopic object like you is made up of a gazillion pieces, all of them as exposed to their quantum behaviour as a single electron or a single photon. But when you start averaging over so many interacting parts, the prosaic story of our everyday existence begins to emerge and the quantum fuzziness is much harder to spot.

I imagine you are feeling a little uncertain about all of this. Excellent! That's exactly how you should be feeling. You see, uncertainty is at the very heart of quantum mechanics. In fact, quantum mechanics will just break into a million pieces if you don't assume something called the *uncertainty principle*. This says that you cannot have precise knowledge of both the position and the momentum of an electron, or indeed any other particle. Quantum mechanics forbids it.

To understand why, imagine you have a high-resolution microscope that can pick out an individual electron and work out where it is. The trouble is that you have to shine some light on the electron to see it. The beam of photons carries momentum and when it strikes the electron it passes on some of that momentum. We can't really be sure how much. To reduce this uncertainty, we need a much lighter touch. First, we need to dim the beam so that we are firing only one photon at a time. But even that is not gentle enough. We also need to reduce the momentum of the individual photons. But now we must remember what de Broglie taught us: low-momentum photons have a really long wavelength. The problem is that the resolution of the microscope depends on the wavelength of the incoming light – the longer the wavelength, the worse the resolution. If you want to be really certain about the momentum of the electron, you have to be really *un*certain about its position.

This analogy is due to Heisenberg himself, the proud Bavarian who discovered the uncertainty principle in 1927, at the height of the quantum revolution. The analogy is a little loose since it doesn't really take into account the quantum nature of the interaction between the electron and the photon. To properly understand the uncertainty principle, we need to express it in the right way. Whenever you try to measure the position of an electron, the best you can do is pin it down to some fat region of space of width Δx. The same is true of the momentum. You can't be certain of its value – you just know it's in some range of momenta, spread across a width Δp. Δx and Δp are often said to be the uncertainty in the position and the momentum, respectively.

Heisenberg's principle says that together they must obey the following rule:

$$\Delta x \Delta p \geq \frac{\hbar}{2}$$

If you want precise knowledge of the electron's position, the fat region of uncertainty Δx must shrink to zero size. Likewise, to have precise knowledge of its momentum, Δp needs to vanish. Heisenberg's rule tells us that both these things cannot happen at once. If you want better knowledge of the position, you have to give up knowledge of the momentum, and vice versa.

There is another version of the uncertainty principle, this time related to the uncertainty in the *energy* of a particle, ΔE, and the uncertainty in its *time*, Δt. It is the extra ingredient you need if you want to talk about uncertainty in spacetime, as Usain Bolt might be inclined to do. It takes on a very similar form:

$$\Delta E \Delta t \geq \frac{\hbar}{2}$$

The best way to make sense of this particular formula is through the sound of music. This is because uncertainty is really a property of waves – you don't just see it in the probability waves of quantum theory but in the sound waves produced by a musical instrument. My friend and colleague Phil Moriarty covers this at length in his book *When the Uncertainty Principle goes to 11*. Phil likes to play electric guitar. Suppose he plucks at the A string, letting the note ring out for as long as possible. The note hangs there in the air for several seconds until the energy has dissipated away. He knows as well as anyone that this particular noise comes from a sharp combination of sound waves of different frequency. If you were to look at the frequency spectrum more carefully, you would see a series of narrow peaks, picking out the individual harmonics for that particular string.

Because Phil is a metalhead, he also likes to 'chug' on his guitar, pushing the ball of his hand down on to the bridge to dampen the note. The result is the classic heavy-metal sound – the same note as before, only now it is delivered with a characteristic thud. If you analysed the spectrum of the chug, you would find the same harmonics as before (after all, it is the same note), but the peaks would have melted into one another, forming an ugly amorphous blob of indefinite frequency.

The amplitude of Phil's first note plotted against frequency (top) and time (bottom). The note corresponds to series of very narrow-frequency bands and continues to sound for quite some time.

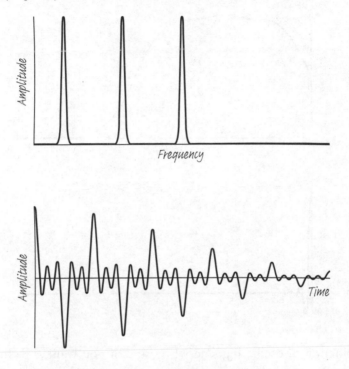

The difference between these two guitar sounds is at the very core of the uncertainty principle. The first sound is precise in frequency, as seen by the narrow peaks in its spectrum. But it is imprecise in time – the note lasts for so long we cannot be very specific about when it actually happened. For the chug, it is the other way around – precise in time, thanks to the brevity of the sound, but imprecise in frequency. What we are seeing is a trade-off in each case: precision in frequency versus precision in time.

It's the same for probability waves. To connect with the uncertainty principle, all we need to do is pass from frequencies to energies, using Planck's currency converter, $E = \hbar\omega$. At the end of the day, the uncertainty principle is nothing more than the elementary mathematics of Frenchman Joseph Fourier, dating back to the early part of the nineteenth century. Fourier showed how any signal could be built from a

The amplitude for Phil's chug plotted against frequency (top) and time (bottom). This time the note doesn't last very long at all and the frequencies are spread across a much broader range.

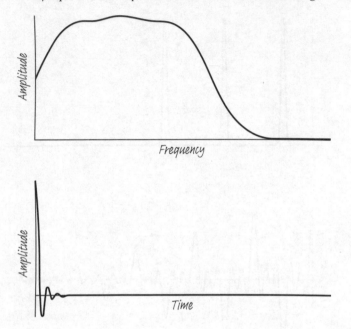

combination of oscillating sine waves and, if you wanted to localize the signal – to pin down its location in time or space – you would need lots of waves cancelling one another out in lots of places. For an electron or a photon, if you want to know where it is, you need a single sharp peak in its probability wave. Fourier tells us that this means you need lots of waves oscillating over many different wavelengths superimposed on top of one another, cancelling each other out everywhere except in the vicinity of the particle.

There is an important aspect to the quantum story that we've been avoiding, at least until now, because it is arguably the most disturbing. Do you remember when you were chasing the fugitive in the shopping mall? You weren't really sure where they were hiding, then all of a sudden, one of your police officers caught them and you knew exactly where they were. In an instant, you went from a wave of probability spread right across the mall to a sharply localized peak at the point of their capture. What was the physics that described that transition?

We face the same question when we detect an electron – according to Bohr, the wave function just instantaneously collapses into one place or another the moment you carry out a measurement. You cannot describe that with an equation like Schrödinger's – so how do you make sense of it? When I was a student at Cambridge I asked my tutor about this. He replied that he had asked the same question of the great quantum pioneer Paul Dirac – who confessed to being completely non-plussed. But I was a student a long time ago. Nowadays we know a lot more (if not everything) about what actually happens but, in order to explain it, I first need to tell you the story of Schrödinger's dog.

In a daring raid on Buckingham Palace one of Queen Elizabeth's favourite corgis has been captured by a radical group of science teachers calling themselves The Disciples of Schrödinger. Their goal is to teach the public about science using whatever it takes to get their attention. Soon after the raid, they post a video online showing the dog being locked inside a large box. The box is sealed completely so that no one can see or hear what is going on inside. The disciples assure the viewers that there is enough air for the dog to survive for at least two hours. But they also warn of something else. Alongside the corgi is a small radioactive device. In a one-hour spell it has a 50–50 chance that one of its atoms will decay. If that happens, it will trigger a chain of events so that a gun will fire, killing the dog instantaneously. However, there is also a 50–50 chance that no atoms decay and the corgi survives. The video cuts to a live feed. The corgi is still in the box. The disciples reveal that it has been in there for nearly an hour, and they invite viewers to speculate as to what state it might be in. Is it alive or dead? Social media is flooded with reaction:

> I have a really bad feeling about this. #dogisdead
> Everybody needs to stay positive. #dogisalive
> The dog is alive, and it is dead. #superposition

And then they open the box. They sombrely declare that the corgi has died. Or perhaps, like the queen, you'd prefer another ending, where the dog survives. It really doesn't matter. The point is that when the disciples open up the box and look inside, the dog is alive or it is dead. There is no other way to end the story.

But what about that moment when the disciples asked the question, just *before* they looked inside the box. What then? Well, like everything else in quantum mechanics, the corgi should be described by a probability wave. One such wave will describe a dog that is alive; another wave will describe a dog that is dead. When it first enters the box, the dog is barking relentlessly and seems determined to bite a chunk out of someone. It is clearly alive and should be described by the first of our probability waves – the living wave. However, as time goes by, the dog's wave evolves into something more exotic – ripples of a dead dog superimposing themselves on to the ripples of a live one. The probability of the corgi being dead or alive ends up stretched across both possibilities, just like it was stretched across the position of the fugitive in the mall. So before they look inside the box – before anyone performs a measurement – it would seem that the corgi is both alive *and* dead.

Hashtag superposition.

Now that's all well and good, but when the disciples finally open the box and take a look they only ever see a living dog or a dead one. They never see a bit of both. It seems as if the corgi's wave has collapsed into a state of alive or dead, just like it collapsed for the fugitive when they were caught in the middle of the shopping mall. If the dog really is both alive and dead, why do the disciples never get to see as much? Why don't they see any quantum fuzziness? To make sense of that we need to think about everything that surrounds the dog – from the disciples to the watching world to all of the air molecules that fill the box in which it is held. Let's call all this the *environment*.

When the environment comes into contact with the dog, it begins to interact with it, billions of atoms and photons continuously bouncing this way and that, trading energy, momentum and whatever else they have to offer. But here is the thing: superpositions are contagious. As soon as it makes first contact, the environment sees a superposition of dogs. Should it react to the dead dog or the one that is living? In the end, it cannot choose so it doubles up and reacts to both. This duplicitous behaviour is a symptom of a new and improved superposition: in one half of the superposition we identify a sad environment tangled up with a dead corgi, unable to separate from it; in the other we see a happy environment tangled up with a living corgi.

Anyone or anything that is part of the environment will be able to see only what the environment allows them to see. What would we need for the disciples to see a dog that is alive *and* dead? We would certainly need a superposition: a wave of probability that picks out the chance of a happy environment with a dog that survives; and a wave of probability that picks out the chance of a sad environment with a dog that dies. However, to feel the quantum fuzziness we would also need the waves to overlap – we would need happy and sad to interfere with one another, just like the states of the electron passing through the slits in Davisson and Germer's classic experiment. It seems as if all the ingredients are in place. After all, I've just told you how the environment is press-ganged into joining the superposition, so the superposition is certainly there. Why then do the disciples never see a dog that is both dead and alive? The trouble is that the environment is big; and the bigger it gets, the more the happy wave is torn apart from the sad and the less they overlap. This process is known as *decoherence*. As more and more of the environment comes into contact with the corgi, directly and indirectly, the probability waves that describe it are pushed further and further away from one another. The happy wave and the sad wave can no longer interfere in any meaningful way and the quantum properties of the dog are heavily disguised. Decoherence happens so quickly that when a disciple checks on the corgi they are virtually guaranteed to see it as either dead *or* alive. They never see both.

Although this explains why we don't see a quantum blur in our day-to-day lives, it doesn't really explain away the question that left Dirac so nonplussed. By the end of the process the dog and the environment are still in a superposition, albeit with barely any overlap. One school of thought states that the puzzle is of our own making, a symptom of our desperate need to cling on to determinism. There is a risk in attaching too much *reality* to the wave function, as Schrödinger and many others were inclined to do. The wave function is not really something you can *grab* hold of. Rather, you should think of it as the guardian of probability. Its role is to give you a flavour of what might happen in an experiment, just as a set of odds gives you some idea of what might happen in a horse race. The outcome of the experiment or the race is the outcome of the experiment or the race. It is what it is. What is there to worry about?

There is another important element to the corgi's story that we'll need to understand when we finally return to the question of doppelgängers (remember them?). We now know that the corgi and the environment end up as a superposition of tangled-up states. This state is an example of a *pure* state. Even though it is very complicated it still behaves like a wave and contains complete information about the true quantum state of the dog and the environment in which it finds itself. However, in reality, for large systems, we never know the pure state exactly. Keeping track of that amount of quantum information is not just impractical, it can sometimes be impossible, especially when there are black holes around destroying the records of its prisoners. The way to handle this is to resurrect the spirit of Boltzmann – we need to take averages.

In the story above, the queen's main concern is the wellbeing of her beloved pet. She has no interest in the exact state of some of its atoms, the air molecules that surround it or the box in which it is held. And she certainly doesn't care about the state of a radical band of science teachers who have taken her dog captive. To describe the quantum state of the health of the dog and *just the health of the dog*, she needs to trade in her ignorance of everything else and take averages. The way to do this is to take all possible environments, tangled up as they are with all possible states of her beloved corgi, and calculate their average contribution. What is left at the end? What she gets is a so-called *mixed* state. It is basically a *list* of possible states linked to the corgi's wellbeing (e.g. the state of a dead dog, or the state of a living one) with some associated probabilities. These probabilities give her an idea of what someone might see when they finally look inside the box.

Now these mixed states might not sound a whole lot different to the pure states we have also been talking about, but they are *not* the same thing. A pure state is a genuine wave – a *superposition*, one ripple superimposed on another, giving you another, more complicated wave, but another wave nonetheless. A mixed state is just a list, not a superposition. It does not behave like a wave. When we think of a pure state describing the dog *and* the environment, there are certainly superpositions where we can think of the dog as both dead and alive. However, when we start thinking of a mixed state describing *just a dog*, we cannot really say if it is dead or alive or indeed any combination of the two. This is because we have absolutely no idea. We can give a list of some particular pure states we think it *could* be in and the associated probabilities, but that is the best we can do.

One way to understand this better is to imagine yourself listening to the classic Beatles track 'Let It Be'. You are listening through a set of headphones and they are set up so that one ear plays the spine-tingling piano instrumental while the other plays the sound of Paul McCartney singing a capella, whispering words of wisdom with characteristic charm. If you wear both ears at once, you will of course hear a superposition of both sounds and get to enjoy the full tune as it appeared in the charts in 1969. Each of these sounds can be thought of as a pure state: the piano instrumental, McCartney singing, as well as the glorious combination of the two. All three of them are a superposition of waves, albeit sound waves rather than probability waves, as in quantum mechanics.

Now imagine another scenario in which you have accidentally broken the headphones so that one of the ears isn't working. You can't be sure which one was meant to play which sound, so until you have a listen you don't know which of the two sounds will be missing. You have lost some information. What you now have is a mixed state: a *list* of two pure states – the piano instrumental and McCartney singing – each with a 50–50 chance of being played through the surviving ear.

A pure state tells you everything there is to know about a quantum system. If you like, it is complete quantum information. Of course, that doesn't translate into absolute predictions about the outcomes of an experiment. Those are still shrouded in probability because, in quantum mechanics, a pure state happens to be a wave of probability and the best it can ever do is tell you where an electron is *likely* to show up, not where it *will* show up. On the other hand, with a mixed state, there is quantum information that is actually missing. We can't even be sure which particular superposition describes the system because that knowledge has become tangled up with an unknowable environment. If all we care about is whether a dog is dead or alive, then there is a load of junk information with which we don't need to concern ourselves. Our knowledge is incomplete, but so what?

The mixed state gives us an idea of what to expect when we perform the measurements that matter to us.

I've taken you on a difficult quantum journey, deep into the microscopic world of probability and uncertainty, but I promise it was more than a curious sideshow. It was essential in our quest to discover your doppelgänger, to understand who they are; to understand who you are. We now know that you should not be identified by a particular arrangement of atoms because such a description is impossible. It would require us to know the exact position and momentum of all the particles in a human body, something which Heisenberg has told us is forbidden by quantum law. The truth is we should think of you as a complicated quantum state, governed by waves of probability superimposed on top of one another. But to compare with your doppelgänger, do we really need to know everything about this complicated state? Do we need you to be pure?

WHERE IS YOUR DOPPELGÄNGER?

What are *you*? What does it mean to be the same as you? My brother, Ramón, and I share a lot of DNA. We both like Stiff Little Fingers and we both support Liverpool Football Club. If that was all we cared about, we would be doppelgängers. But we are different in a lot of other ways: for example, my neck is freakishly long whereas his is a normal size. For true doppelgängers, we should not tolerate *any* differences, although, as we will see, that could be a very dangerous game indeed.

Compared to an electron, there is a lot of you. That's to be expected. To construct something as sophisticated as the person reading this book, you need a great many things: quarks and gluons tied up in protons and neutrons; atomic nuclei shrouded by probabilistic clouds of electrons; atoms bonded together in complex molecular chains; then a trillion of these molecules gathered carefully together in a trillion or more cells. And to make it even more complicated, all of that stuff is tangled up with the world that surrounds it. When you were at school, do you remember how a little bit of gossip would spread through the classroom? It would often take the form of a handwritten note, perhaps something along the lines of 'Degsy is gonna ask out Helen

Jones – <u>pass it on</u>.' The last bit was always underlined to remove any doubt as to the importance of the instruction. As the note meandered its way around, different people would react in different ways to the news – jealousy, excitement, indifference. These reactions would often trigger a new set of reactions and interactions. Whatever it did, one thing was certain – in no time at all the entire class would be tangled up as one with knowledge of Degsy's intentions. It is the same with you and the observable universe. The universe has been passing notes around since the beginning of time and it is tangled up with every single piece of you. That is an awful lot of information to keep track of.

When someone looks at you or even asks what you are thinking about, they are obviously not acquiring all the available information. The truth is they don't care whether one of the electrons deep inside your small intestine happens to be spin up or spin down.* Whenever we talk about a person (or an egg or a dinosaur or a gas of particles) we never really think of them as pure, because there is an awful lot of detail we don't need to know. You are no exception – you are not pure. You are mixed. All we would ever really do to describe you is specify a list of states – the microstates – and their associated probabilities. But what should we make of the missing information? And what would it take to find it out?

What we don't know is hidden in the list of probabilities. We cannot do better – not without measurement. For example, some of the microstates that describe you have the electron in your intestine spinning upwards with some probability, while others have it spinning down with some other probability. Don't be fooled into thinking that the electron is really spinning upwards and you just don't happen to know about it. In quantum mechanics, there is no absolute truth – again, not without measurement. Until you set up

* There is an intuitive notion of spin from our everyday world – it is the momentum you associate with orbital motion, when things are rotating in some way. In quantum mechanics, however, there are actually two types of spin. There is the quantum version of orbital spin, but there is also a new intrinsic sort of spin. This new spin is similar to its orbital cousin but has no analogue in our everyday world. For electrons, we can measure the intrinsic spin to be either 'up' or 'down'. This can be done using a time-varying magnetic field, as in the Stern–Gerlach experiment, first performed by Otto Stern and Walter Gerlach in 1922.

a miniature Stern–Gerlach experiment in your small intestine and perform a measurement on the spin of the electron, all we can ever talk about are the probabilities of spin up or spin down. This logic applies to each and every one of the great many things we could ask about you, right down to the microscopic level. Unless you perform all the appropriate measurements you need to accept who you really are – you are a quantum schizophrenic, a vast family of microscopically distinct versions of yourself, all of which are as real as any other.

The only way to be cured of this schizophrenia is to perform more measurements. It is the *only* path to purity. The trouble is that this requires us to perform a vast number of measurements – there are more than a billion billion billion atoms in your body, and you would have to deconstruct the structure of every last one of them. That kind of experimental invasion will almost certainly destroy you. It is hard to imagine how we can probe all of your microscopic structure without exposing you to the kind of energies that would tear your atoms apart. In the end, we cannot escape the fact that performing the experiment will affect what you are – and chances are you will end up as plasma. Sometimes it is better not to know.

But suppose we *can* perform all the necessary measurements without obliterating you – what then? Well, then you really would be one part in a doppelgängion. You would be one of $10^{10^{68}}$ possible microstates, perfectly pure, albeit only for an instant. The clever team of experimentalists who successfully recorded all of your microscopic structure could now begin the search for your doppelgänger. Of course, it is a lot of information for them to carry around. As we will come to understand in the next chapter, they would be well advised to store it in a physically large enough space – larger than a human – to avoid it collapsing into a black hole. But assuming all the relevant safety procedures are carried out, the search can begin. The team starts with the cubic metre of space to your right and they perform the necessary measurements. Do they yield exactly the same results as they did for you? Almost certainly not, so they move on to the next cubic metre, then the next, and they carry on for as long as they dare. The chances of recovering the same results in any one experiment are slim – a doppelgängion to one chance – but do something often enough and the unexpected can occasionally happen. It is why you shouldn't be that surprised that Leicester City won the Premier League in 2016. If the doppelgänger search team travel to doppelgängian distances and perform their measurements a doppelgängion number of times, they have a chance of overwhelming the odds. Don't be surprised if they find another you sitting there reading this book.

Come on, seriously?

I expected you and your other self to react this way. But consider this: doppelgängian distances are tiny compared to the *googolplician* universe. As fractions go, the ratio of a doppelgängion to a googolplex is imperceptibly small. This means that the slim odds of finding your doppelgänger are not just overwhelmed, they are overwhelmed again and again. It is even more remarkable when you realize that a doppelgängian distance is almost certainly an *over*estimate – it was derived by demanding an exact match between you and your doppelgänger, something which is in serious danger of killing you both. A safer and more relaxed definition would mean that doppelgängers are likely to show up at even *shorter* distances. So, as rare and sophisticated as you might be, and however strict your matching criteria, in a *googolplician* universe it would be utterly implausible to deny your doppelgänger. It would be implausible to deny a multitude of doppelgängers.

If the universe is big enough, your doppelgänger is out there.

So is it big enough? Well, we have to be clear about what we mean by *universe*. First of all, there is the *observable* universe. If the universe had a beginning, then light from the most distant worlds would not necessarily have enough time to reach us – there would be a limit to how far we could see. And we know for a fact that the universe had a beginning – you can tell as much by looking up at the night sky. What do you see? Save for the romantic twinkle of a handful of stars and planets, you see an inky blackness. But this is *not* what you would see in an infinite universe that had existed for ever. The sky at night would be as bright as day – in every direction you looked you would see the light from a star, young, old or unimaginably ancient. It was the German astronomer Heinrich Olbers who first pointed this out. He imagined this infinitely vast universe, unchanging in time, with stars spread evenly throughout. With an eternity to reach you, there would be no limit on the age of the stars you would see. The more distant stars would appear dimmer, of course, but they would also be more abundant, and everywhere you looked, you would see them. In Olbers' universe, night becomes day.

But night is not day, and that is because the universe is constantly reinventing itself. As time goes by, the gaps between stars and galaxies are getting bigger and bigger, not because they are trying to run away from each other but because space itself is growing. It is literally *expanding*. Wind the clock back and the universe will shrink and, at some point it will shrink into nothing. This is the beginning of the universe, perhaps the most significant date in history, some 14 billion years ago.

There are various ways to measure the age of the universe. One way is to capture the light from the most explosive and violent deaths in the furthest reaches of what we can see. These are the distant supernovae, the beacons of dying stars far, far away. We assume that these distant supernovae are more or less the same as the ones nearby, and by comparing the properties of the light we receive we can read off valuable information about the history of the universe. Another way to measure the age of the universe is through the cosmic microwave background (CMB), a stream of radiation that has journeyed through space ever since the first atoms formed. These two age measurements are in slight tension with one another, not drastically so, but enough to start getting people very excited about a hint of new physics. That said, as a ballpark figure, both are broadly consistent with a universe that is roughly 14 billion years of age. The important point for us is that this age is finite, and that puts an upper limit on the distance light could have travelled since the beginning of time. You might think it equates to a distance of around 14 billion light years, but that would ignore the expansion of space. In the end, it turns out that the furthest reaches of the observable universe are around 47 billion light years away. Anything beyond that is too far away to communicate with us – there are no signals, light or otherwise, that could have reached us since time began.

So how does 47 billion light years compare to a googolplician universe?

It doesn't.

It is completely insignificant. Despite what my cousin once said, there is really no hope of finding your doppelgänger within the observable realm. But what about beyond? How far is the realm of existence? Does it extend beyond the imaginary wall at 47 billion light years? Are there *wildlings* beyond the wall? And what would it even mean for the universe to stop?

The universe certainly doesn't stop at 47 billion light years. It extends much further than that, into regions that are hidden from terrestrial view. You could even travel to these distant realms – if you could live long enough, that is. The universe *could* eventually stop, coming back on itself, wrapping itself up like the surface of a tremendous sphere. You could even imagine a cosmic Magellan on the ultimate expedition to circumnavigate the entire universe. If the universe really is a tremendous traversable sphere, the CMB photons could potentially tell us. If it is a sphere, it must be undetectably large, with a diameter of at least 23 trillion light years. That means the entire universe is at least 250 times larger than the portion we are able to see. These hidden depths are large, but are they large enough? The universe may reach beyond 23 trillion light years, but does it reach as far as a googolplex?

To glimpse the true size of the universe, we need to return to its childhood. Children like puzzles, and in the CMB there is a puzzle. If you happen to be floating aboard the International Space Station and you look to your left, there will be CMB photons that strike you square in the face. This is radiation that has cooled on its triumphant journey across the age of the universe, down to an average temperature of just 2.7 kelvin. Now look to your right. You are struck by another stream of CMB photons, and they too have an average temperature of 2.7 kelvin. Whichever way you look, the CMB photons are at this temperature. You may not think that is strange, but it really is. These photons bring with them knowledge of the worlds they came from, and they are all saying the same thing. This can only mean that the distant worlds know something of each other, but how can that be? After all, when the photons began their journeys, these early worlds were unobservable to one another. No signal could ever have passed between them. How did they conspire to spread knowledge of one temperature right across the CMB sky? It is a bit like stumbling across a tribe in the deepest part of the Amazon: although they have never had any contact with the outside world, for some reason, they converse in perfect English. Despite all that you had been told, you would no doubt suspect that, at some point in their history, they had met an English person.

And so it must be that the distant realms at opposite ends of the CMB sky have met at some point in the past; at some point, they have communicated. But if they are too far apart to have traded signals, how did they do this? It is possible that the child universe came up with an ingeniously simple solution to solve this conundrum, through a process known as *inflation*. Inflation suggests that the two distant realms were once very close: neighbours, trading signals, communicating information. And then suddenly they were blasted apart, torn away from each other faster than light. In a way it is tragic, but it is also strange. How could they be blasted apart faster than light? It is certainly true that nothing in space can outrun light itself – not even Usain Bolt. But that is not what is happening here. It is *space* itself that is growing faster, pushed violently apart by a curious little devil known as the *inflaton*. We don't know very much about the inflaton. We think it might be a bit like the celebrated Higgs boson we'll encounter later on; it might even be the Higgs boson wearing a very different hat at a very different time – but we don't know for sure. We don't even know if there is one inflaton or two. Whatever it is, we know that it rapidly created so much space between neighbouring worlds that by the time it was done they had lost all ability to communicate. But the important thing is that they still remembered each other, and they passed that information on to the CMB photons. That is why they have roughly the same temperature.

We are finally led to a googolplician universe once we start asking about the beginning of inflation – why did it start in the way it did? The answer may lie in something known as *eternal* inflation, a never-ending process of universe creation. The idea is that the inflaton spent its time randomly trying on different looks. It hopped, quantum mechanically, from one value to another. For the most part it didn't do anything particularly interesting until suddenly, in the tiniest corner of the baby universe, it hopped on to just the right value to trigger the blast. Like the seed of a great sequoia, the little corner would have grown into something gargantuan – all of the universe that we see. But here is the nub. The inflaton continued to hop, bouncing at random from one value to another, and it did this at every single point in space. Every so often, somewhere, in some tiny forgotten corner of the universe, it bounced into the sweet spot and *boom*! A gargantuan of space was born. And then it happened again. And again. And the more it created, the more it chanced to happen. A mighty leviathan grew and grew to monstrous proportions until, eventually, it dwarfed even a googolplician universe. And as the cosmic Magellan would testify as he travels towards the most far-away realms imaginable, in a universe this large . . . there be doppelgängers.

My cousin Gerard was right.

Graham's Number

BLACK HOLE HEAD DEATH

When I was a kid there was a popular TV show on the BBC called *Think of a Number*. The presenter, Johnny Ball, would bounce around the stage armed with a magnificent array of props and costumes, filling our young, impressionable minds with the joys of science and mathematics. Of course, I absolutely loved it. Think of a number – harmless educational fun. Or was it?

It's perfectly OK to think about seven or fifteen or four hundred and seventy-six thousand five hundred and twenty-two. But what happens when you think about Graham's number? Well, if you do that, it's really not OK. If you think about Graham's number in the wrong way, you die. Looking back, Johnny Ball really ought to have named his show *Think of a number that won't kill you*, but I guess health and safety was less of an issue in 1980s Britain.

It is tempting to compare death by Graham's number to the plight of some people caught up in the eruption of Mount Vesuvius in AD 79. You may have seen the images of the victims in Pompeii: their final moment turned to stone, killed by a blast of intense heat from the pyroclastic surge and preserved for ever in a tomb of ash. These were the lucky ones. In nearby Herculaneum and Oplontis, there is grim evidence for an even more disturbing end: remnants of shattered skulls blasted apart by the rapid boiling of cerebral fluids after the volcano's eruption. These victims died because their heads exploded.

Graham's number can result in brain trauma that is even more spectacular. It can happen if you are forced to think about it digit by digit – if its decimal representation is thrust unceremoniously into your imagination. For a while you feel nothing untoward, a string of digits growing larger and larger in your mind's eye. And then it happens.

Black hole head death.

The truth is that you cannot think about Graham's number, at least not in all its gargantuan glory. There is simply too much of it for you or anyone else to contend with. It is not a question of intelligence but of physics. If you try to cram that much information into a human head, that head will inevitably collapse to form a black hole. As we will see, black holes put a limit on how much information you can squeeze into any particular volume of space, and your head is nowhere near big enough to cope with all the information contained in Graham's number. This is the problem with it. It's not just big, it is magnificent, far larger than a googol or a googolplex or even a googolplexian. Graham's number and all its digits cannot exist in your head, or in the observable universe, or even in a googolplician universe. There is just too much information contained in that decimal representation to squeeze it all in.

Why would anyone invent a number that is capable of killing you? The man responsible was the award-winning mathematician Ron Graham. Graham did not respect the mathematical stereotype. As a baby-faced fifteen-year-old accelerated to college in Chicago in the early 1950s, he took up trampolining and juggling and was soon good enough to perform with a circus group known as the Bouncing Baers. Even in old age, he continued to bounce, albeit in the comfort of his own home. As his friends report, with Ron Graham, you came to expect the unexpected. In one moment, he would be discussing mathematics, and in the next he would flip into a handstand or dance around you on a pogo stick.

The story of Graham's number really begins in the early part of the twentieth century, with another colourful mathematician by the name of Frank Ramsey. Ramsey was something of a polymath and a member of a secret society of intellectuals known as the Cambridge Apostles.* He was a student of the great economist John Maynard

* The Cambridge Apostles are essentially an intellectual discussion group comprised mainly of Cambridge University undergraduates. It gained some notoriety in the 1950s and 1960s when two of its former members, Guy Burgess and Anthony Blunt, were exposed as part of the Cambridge spy ring passing British secrets to the Soviet Union. Ramsey was an Apostle in the 1920s, in the decade before Burgess and Blunt were elected and, although he was

Keynes who later recommended him as a fellow of King's College, my old college at Cambridge University. As undergraduates we knew all about Keynes – I lived in a building named after him – but nobody ever spoke about Ramsey. They should have done. Ramsey died in 1930 at just 26 years of age from chronic liver problems, having already made major contributions in mathematics, economics and philosophy. However, his greatest contribution happened almost by accident via a small and incidental theorem buried deep within his 1928 paper on formal logic. That theorem was the seed for a new branch of combinatorial mathematics that now bears his name.

Ramsey Theory is concerned with deriving order from chaos. It's a bit like watching parliamentarians discussing Brexit and asking yourself: among all this disorder – this cacophony of ego and opinion – can I find some pockets of agreement, some commonality? I could ask the same question when I host a dinner party.[1] Imagine an event to which I've invited six very different guests, friends and family from every corner of my life, with very different backgrounds, experiences and opinions. I sit them all around the table and, as a good host, I try to establish who knows who. Algernon knows my daughter, Bella. He is an old friend from university and has been up to the family home to visit from time to time. Algernon now works in the music industry. He likes to remind people that when he worked in a record shop Leo Sayer came in and bought a dozen Leo Sayer CDs (true story). Bella is still at school but she is hoping to become an artist one day. Algernon also knows Clarkey from their time together at university. Clarkey works as a sports broadcaster. He doesn't know Bella because he tends to avoid children whenever possible. I make a note of all of this with the following diagram:

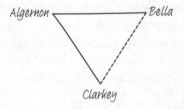

---

politically to the left, there is certainly no reason to suggest he would have been involved in espionage.

The solid lines obviously indicate two people who know each other, whereas the dashed lines show that they don't. Next, I factor in Deano, a professor at an Ivy League university. He also went to university with me, Algernon and Clarkey but, like Clarkey, he doesn't really know Bella. I update the diagram:

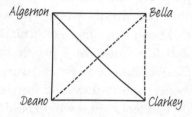

Now I throw in the last two guests. Ernest and Fonsi both know Bella, but they don't know each other, or indeed anyone else at the party. Ernest is an engineer whose grandfather was responsible for bringing grey squirrels into Britain from North America (again, a true story). Fonsi is an aspiring politician. Once again, I update the diagram.

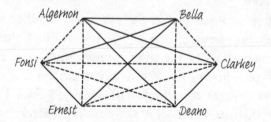

Already, with just six guests, this network of lines, solid and dashed, has started to look chaotic. However, if you look within the chaos you will start to see a degree of order. For example, Algernon, Clarkey and Deano form a clique – a group of three people all of whom know one another. Clarkey, Ernest and Fonsi form another sort of clique – a group of three complete strangers. You will notice, however, that there are no cliques of four of any sort.

Networks like these are at the heart of Ramsey Theory. It is no surprise that we found ordered cliques of three at a dinner party for six. That was guaranteed. In fact, six is the minimum number of people you need to make sure that it happens. However, as we have seen, six guests are not enough to guarantee cliques of four at a dinner party. It turns out that you need at least eighteen guests to guarantee that. These are the Ramsey numbers. Simplifying the jargon,[2] we might say that the third Ramsey number is six, while the fourth is eighteen.

Ramsey showed that you can get cliques of any finite size as long as you invite enough people to the party. But he didn't always know how many people that should be. Already when you start to look for cliques of five people, things get much more difficult. Most mathematicians think you need at least forty-three people at your party to guarantee a clique of five, but no one knows for sure. The minimum number is somewhere between forty-three and forty-eight.

To pin it down exactly, mathematicians would need to reproduce all possible networks and then poke around to see when cliques of five are guaranteed to occur. You could try to do this on a computer, but there simply isn't enough computing power to do it properly. When you have forty-three guests, it turns out that you are asking the computer to scan its way through 2^{903} different networks. That is considerably more than a googol. Even the supercomputers of today baulk at those sorts of numbers.

To guarantee a clique of six, the minimum number you need to invite is somewhere between 102 and 165. Obviously, the problem of finding an exact value for the sixth Ramsey number is considerably more difficult than finding the fifth. This was emphasized in apocalyptic terms by the great travelling mathematician Paul Erdös. Erdös imagined an alien invasion – an army of extraterrestrials far more advanced than us – landing on Earth and demanding that we tell them the fifth Ramsey number or face annihilation for our stupidity. In this case, Erdös's strategy was to combine the power of all the world's computers and trust in mathematicians to provide the answer. If, however, the aliens were to demand the *sixth* Ramsey number – well, then there would have been no point. We should figure out a way to annihilate the aliens before they annihilate us.

Erdös's colourful expression offers a glimpse into his unique character. Born in Budapest on the brink of the First World War, he was an eccentric who spent most of his adult life on the road, rarely settling in one place for more than a month. He travelled continuously, from one collaborator to another, across continents, searching for new solutions to his compendium of mathematical problems. If Erdös arrived at your door with his suitcase, he would expect you to find him a bed for as long as he wanted, to feed him and to plan and organize his affairs. If you had children, he would refer to them as epsilons, a nod to the notation mathematicians use when they want to describe something that is infinitesimally small. He would also have a problem for you to solve. That was his greatest skill – to connect a mathematical problem to the one person who could help him solve it. Throughout his remarkably unusual career, fuelled by an addiction to amphetamine, Erdös wrote over 1,500 papers and had over 500 collaborators. His reach in mathematics is so great that academics now talk about their Erdös number, since most of us can connect to him via a small handful of collaborations.

Ron Graham has an Erdös number of one. He and Erdös were very close, so much so that Graham set up an 'Erdös room' in his home where Erdös could stay during his visits and store things while he was away. Graham even took care of Erdös's finances, collecting his cheques and paying his bills. But it wasn't through Erdös that Graham stumbled upon his now famous number. That was through a collaboration with a fellow American mathematician, Bruce Lee Rothschild, and later, with Martin Gardner, a columnist for *Scientific American*.

Graham and Rothschild were interested in a very particular problem in Ramsey Theory. To understand it, let us add two more guests to the party: Graham and Harold. Graham is Bella's uncle, while Harold is

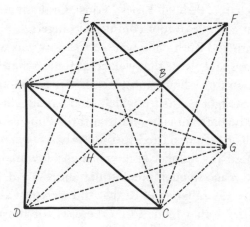

something of a mystery. It seems he is fluent in five different languages, but no one really knows who he is or what he does, and he won't give much away – perhaps he is a spy. It doesn't really matter. What is import- ant is that we now have eight guests, which means we can imagine placing them at the vertices of a cube to create a new kind of network.

Suppose I decide to cut a slice through the network. For example, I could slice along the diagonal, through Bella, Clarkey, Ernest and Harold. These four form a kind of subnetwork, and one that is much easier to draw on a flat piece of paper:

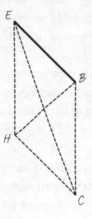

But it isn't any sort of clique – it is an unremarkable combination of friends and strangers. Could we have chosen a more interesting slice? In this particular example, the answer is yes – by slicing across the back of the cube, through Ernest, Fonsi, Graham and Harold, we would pick out a clique of four complete strangers.

What Graham and Rothschild wanted to know was whether or not you could *always* find a slice that yielded a clique, whatever the make- up of the cube. In three dimensions, the answer is no – there are other arrangements of eight guests for which there are no cliquey slices. Of course, no mathematician is bound to a three-dimensional world, so Graham and Rothschild started to think about *hyper*cubes in four, five, six or indeed any number of dimensions. How many dimensions do you need to reach to guarantee cliquey slices?

Needless to say, Graham and Rothschild were not able to provide a definite answer – that tends to be the case for most problems in

Ramsey Theory. But they did show that a finite answer existed and were able to place some limits on it: the minimum number of dimensions had to be between six and some gargantuan monstrosity – a finite number beyond anything we could ever hope to comprehend. Contrary to popular belief, the gargantuan upper limit they presented is not what we now refer to as Graham's number. That came six years later, in 1977, when Ron Graham began talking to Martin Gardner. He wanted a simple way to describe the upper limit for Gardner's article in *Scientific American*, so he cooked up something that was slightly more gargantuan. This new number found its way in to the 1980 edition of the *Guinness Book of World Records* as 'the largest number ever used in a mathematical proof'. But it was never actually used for this.

No matter. I want to blow your mind by getting you to think about the value of Graham's number, as he presented it to Gardner. Don't worry. I'm not going to make you think of its decimal representation. Not yet, anyway. For now, we will focus on a much safer way to think about Graham's number, using Knuth's arrow notation, named after the American computer scientist Donald Knuth, who introduced the idea in 1976. Knuth wrote extensively about numbers and computing and was famous for offering a reward of $2.56 to anyone who discovered a mistake in any of his books.* His arrows will provide us with safe passage through the land of large numbers.

We'll start with multiplication: what do we mean when we write 3×4? You probably want to say 'twelve' but let's think a little more about it. What we really mean when we write 3×4 is *three added to itself four times*, or $3 + 3 + 3 + 3$. More abstractly:

$$a \times b = \underbrace{a + a + \cdots + a}_{b \text{ repetitions}}$$

This is just a added to itself b times. Stripping it back like this, we see that multiplication is really just a fancy way of describing *repeated* addition. But what about repeated multiplication?

---

* He chose $2.56 because '256 pennies is one hexadecimal dollar'.

That is what mathematicians call *exponentiation*, normally written using a superscript:

$$a^b = \underbrace{a \times a \times \cdots \times a}_{b \text{ repetitions}}$$

This time we have *a multiplied* by itself on *b* occasions. For example:

$$3^3 = 3 \times 3 \times 3 = 27$$
$$3^4 = 3 \times 3 \times 3 \times 3 = 81$$

You probably call these things powers – my wife calls them 'tuthers' (as in 'to the'). It doesn't really matter what you call them as long as you understand what they really mean. For his part, Donald Knuth has another way of writing powers – he likes to use an arrow:

$$a \uparrow b = \underbrace{a \times a \times \ldots \times a}_{b \text{ repetitions}} = a^b$$

Repeating our examples, that means we have $3 \uparrow 3 = 27$ and $3 \uparrow 4 = 81$.

We could stop at this point, and that is what most normal people do, but we are not normal people. Let's carry on. What do you get when you perform repeated exponentiation? This is called *tetration*. Knuth writes this with a double arrow:

$$a \uparrow\uparrow b = \underbrace{a \uparrow \left(a \uparrow \left(\ldots \uparrow a \right) \right)}_{b \text{ repetitions}}$$

Here *a* is *single arrowed* with itself *b* times. This is also known as a power tower because you can think of it as:

$$a^{a^{a^{\cdot^{\cdot^{\cdot^a}}}}}$$

where the tower of *a*'s has *b* levels.

Let's work out $3 \uparrow\uparrow 3$ and $3 \uparrow\uparrow 4$. These are just power towers of threes, three and four levels high. In other words:

$$3 \uparrow\uparrow 3 = 3 \uparrow (3 \uparrow 3) = 3^{3^3} = 3^{27} = 7,625,597,484,987$$
$$3 \uparrow\uparrow 4 = 3 \uparrow (3 \uparrow (3 \uparrow 3)) = 3^{3^{3^3}} = 3^{7,625,597,484,987}$$

The double arrow allows you to leap from three to 7.6 trillion in one easy step. That's quite an achievement. But Knuth's notation allows you to go even further. All you need to do is use the triple arrow as a repetition of the double arrow:

$$a \uparrow\uparrow\uparrow b = \underbrace{a \uparrow\uparrow \left(a \uparrow\uparrow \left(\cdots \uparrow\uparrow a \right) \right)}_{b \text{ repetitions}}$$

The logic is the same as before, only now a is *double arrowed* with itself b times. The triple arrow is a very powerful beast. Try calculating $3 \uparrow\uparrow\uparrow 3$. This is a triple arrow, so the repetition rule means we need to perform repeated double arrows. For this particular example, we have:

$$3 \uparrow\uparrow\uparrow 3 = 3 \uparrow\uparrow (3 \uparrow\uparrow 3) = 3 \uparrow\uparrow 7{,}625{,}597{,}484{,}987$$

Oh dear. We've ended up with $3 \uparrow\uparrow 7.6$ trillion. That is a power tower –

$$3^{3^{3^{\cdot^{\cdot^{3}}}}}$$

– that is 7.6 trillion levels high! Imagine writing that out in full: if each three were two centimetres high, this tower would stretch all the way to the Sun. For this reason, it is sometimes known as the *Sun tower*. I'll be honest, I'm too afraid to calculate that.

But we are not going to stop there.

How about $3 \uparrow\uparrow\uparrow\uparrow 3$? This is getting really quite silly. If you start to work it out, you will find that you get:

$$3 \uparrow\uparrow\uparrow\uparrow 3 = 3 \uparrow\uparrow\uparrow (3 \uparrow\uparrow\uparrow 3) = 3 \uparrow\uparrow\uparrow (\text{Sun tower}) = \underbrace{3 \uparrow\uparrow (3 \uparrow\uparrow (\cdots \uparrow\uparrow 3)}_{\text{a Sun tower repetitions}}$$

We were already too afraid to calculate the Sun tower and now we have to contend with a Sun tower of *double arrows*. Frankly, this is obscene. A googol and a googolplex have been long since vanquished. We cannot relate to anything of this size. We must accept that we have left the physical realm. And we are still nowhere near Graham's number.

We have no choice but to carry on.

It is at this point that Graham introduces a ladder. Each rung is a number far larger than any that have gone before. The lowest rung on Graham's ladder is usually called g_1 and is taken to be the monstrosity we've just encountered:

$$g_1 = 3 \uparrow\uparrow\uparrow\uparrow 3$$

Take one step up the ladder and suddenly you find yourself hurtling upwards towards

$$g_2 = \underbrace{3 \uparrow \cdots \uparrow 3}_{g_1 \text{ arrows}}$$

Look how many arrows – there are g_1 of them! Four arrows were enough to generate a monstrosity, and now the number of arrows itself has become monstrous. It's a monstrosity of monsters. And we are still nowhere near Graham's number.

Let's take another step up the ladder:

$$g_3 = \underbrace{3 \uparrow \cdots \uparrow 3}_{g_2 \text{ arrows}}$$

There is little point in even trying to describe how big this is. Words have fallen too far behind mathematics. But hopefully you can see the pattern: with each new step on Graham's ladder there is an extraordinary increase in the number of arrows. The effect on the number itself is unfathomable. So, let's just keep on climbing: from g_3 to g_4, from g_4 to g_5, and so on. By the time we reach the sixty-fourth rung we are deep, deep, deep, deep undercover, lost in the land of the largest numbers, unable to recognize who we once were. But we have finally arrived: g_{64} is Graham's number.

It's not very precise, is it? An answer to a mathematical question that is somewhere between 6 and the unspeakably large g_{64}? Ron Graham conceded as much, but to him it highlighted the gap between what you know to be true and what you can actually prove. We know that there is a precise answer to Graham and Rothschild's original question – it is lurking there somewhere in that incredibly large interval – but finding it exactly? Well, good luck with that. In fact, the interval has shrunk a great deal since Graham and Rothschild wrote their paper. We now know that the answer lies somewhere between

13 and $2 \uparrow\uparrow\uparrow 5$. An improvement, but certainly not enough to satisfy the demands of an angry alien race testing humanity with problems in Ramsey Theory.

In the history of mathematics, Graham's number is a true leviathan, but I worry its magnificence is lost in abstraction. To appreciate it more, we shall turn to physics, where we will see why it is a number that is big enough to kill.

TOO MUCH INFORMATION

What makes Graham's number so dangerous? Why will your head collapse into a black hole if you think of its decimal representation? It turns out there is *entropy* in this image of Graham's number – a *lot* of entropy – and whenever you try to cram too much into too small a space, black holes inevitably form. It might sound strange to say that a number can carry entropy in the same way that an egg or a triceratops can, but entropy is intimately related to *information*, and there is certainly information in Graham's number. If I were to tell you its last digit, chances are you'd acquire a little bit of knowledge. If I could tell you its entire decimal representation, your head would try to cram in an awful lot more information. The intake of so much entropy in a confined space would yield only one possible outcome: black hole head death.

To understand the connection between black holes, entropy and the decimal representation of Graham's number, we need to explore the meaning of information. I am thinking of a number, the last digit of Graham's number, and I invite you to figure out what it is. You can ask me any question you like, but I will only give you Yes or No answers. Let's suppose you adopt the following strategy:

> Is it a number from 0 to 4? No
> Is it 5, 6 or 7? Yes
> Is it 5 or 6? No

You realize that the answer must be seven.

You got there in three questions. Your strategy was a good one, narrowing things down by a half with every new question. On average, this strategy will pick out a randomly chosen digit after 3.32 questions.

This is how Claude Shannon, cryptographer and pioneer of information theory, suggested we measure information: the minimum number of Yes or No answers you need to pin down what you want to know.

Shannon combined an obvious talent for computing and mathematics with the practical skills of a prize-winning engineer. He was always making things: from rocket-propelled Frisbees to unicycles to juggling robots. His most mischievous device was a machine that, when you switched it on, released a mechanical hand that promptly turned itself off again. Shannon was also friends with Ron Graham, a friendship that grew from Shannon's interest in juggling: the old man wanted to learn how to juggle and so Graham agreed to teach him. By the end, Shannon was juggling four balls, one more than his robots could manage.

Shannon's interest in information theory stemmed from his wartime work on codes and communication at Bell Telephone Laboratories in New Jersey. He understood the importance of moving information, especially in war, and that doing so could often be difficult or even dangerous. Shannon wanted to work out how to transmit a message effectively even when there is a lot of 'noise' disturbing it, and for that he needed a good measure of its information.

To understand his measure, toss a coin. You only need one Yes or No answer to pin down the outcome of the toss – it is enough to ask: was it heads? So one coin toss carries one bit of information. Five coin tosses will carry five bits; a googol tosses will carry a googol bits. To generalize, we need to relate the number of bits to the number of possible outcomes rather than the number of coins. With five coin tosses there are $2 \times 2 \times 2 \times 2 \times 2 = 32$ different possible outcomes. How do we extract the five bits from thirty-two outcomes? Well, $32 = 2^5$, so the five bits is sitting there in the power of two. When it comes to the last digit of Graham's number, there are *ten* possible outcomes (it could be any digit from 0 to 9). How many bits is that? This is slightly tricky since ten is bigger than 2^3 but less than 2^4, so the answer is somewhere between three and four bits. It turns out that there are about 3.32 bits of information in the last digit of Graham's number.*

---

* 3.32 bits because we have $2^{3.32} \approx 10$ outcomes. Fans of logarithms will note that $\log_2 10 \approx 3.32$.

Of course, Shannon was more interested in words and sentences than in coin tosses. The longest word to appear in a major English dictionary is pneumonoultramicroscopicsilicovolcanoconiosis. It refers to a lung disease caused by the inhalation of volcanic silica in the aftermath of an eruption. Not ideal, but presumably better than an exploding head. What we would like to ask is how much information is contained in the word itself? We could say that each letter is one of twenty-six possible outcomes. As that is somewhere between $16 = 2^4$ and $32 = 2^5$ outcomes, we must have between four and five bits of information per letter. A precise estimate gives 4.7 bits of information per letter.* The entire word contains an impressive forty-five letters, so in total that is 211.5 bits. While this is a reasonable estimate for the total amount of information contained in the word, the truth is that it is an *overestimate*. In English, like any language, there are patterns and rules. For example, consider the word *quicquidlibet*, which literally means *nothing in particular*. Here you twice encounter the letter q and, on both occasions, you know with near certainty that the next letter is going to be a u. How can you say that reading the u gives you 4.7 bits of information when you already knew that it had to be there?

What these subtleties tell us is that there is more to counting information than counting outcomes – you need to factor in probabilities. For example, with a fair coin, if you perform five tosses you will indeed acquire five bits of information. But what if the coin is biased and guaranteed to land on heads? Can you really claim to have gathered any information after seeing it land five heads in a row? Of course not.

Shannon cooked up a better formula for information that takes all this into account. If you toss a coin that has a probability p of landing heads and $q = 1 - p$ of landing tails, Shannon reckoned you would get $-p\log_2 p - q\log_2 q$ bits of information. The formula contains base 2 logarithms because Shannon counted information in terms of *binary* outcomes. Intuitively, it works exactly as you expect it to. For example, if the coin is fair, then $p = q = 0.5$ and the toss reveals one bit of information. If the coin is completely biased towards heads ($p = 1, q = 0$) or

---

* 4.7 bits per letter comes from the fact that we have $2^{4.7} \approx 26$ outcomes. This time, the fans of logarithms note that $\log_2 26 \approx 4.7$.

tails $(p = 0, q = 1)$, then the toss will reveal no information at all. All other possibilities lie between these extremes.

But what about the more complicated things that really interested Shannon, like letters, words, or even sentences? How do we measure their information? Well, suppose you are given the first few letters of an unknown word: CHE. How much information is contained in the next letter once it is revealed? If all letters were equally likely, we would say 4.7 bits. But we know that's not true. Try typing 'CHE' into a message on your mobile phone. What sort of words are popping up in the predictive text? Here are some of the more likely possibilities:

CHEERS
CHEAT
CHECK

This suggests that the letters E, A and C each have a higher probability of appearing than, say, B. If we say that A occurs with probability p_1, B with probability p_2, C with probability p_3, and so on down to Z with probability p_{26}, then Shannon argued that the information contained in the next letter is

$$I = -p_1 \log_2 p_1 - p_2 \log_2 p_2 - p_3 \log_2 p_3 \cdots - p_{26} \log_2 p_{26}$$

As usual, this is measured in bits. Shannon tested the ability of a native English speaker to guess the next letter in a word. His experiments showed that, on average, each character contained between 0.6 and 1.3 bits of information. This might not seem like much, but that is why written English is good for communication. If a letter is missing or input incorrectly, you don't lose too much information and chances are you can styll decode th mxssage.

The most remarkable aspect of Shannon's formula is how closely it resembles another formula developed more than half a century earlier by the quiet man of physics, Josiah Williard Gibbs. We met Gibbs briefly in the chapter 'A Googol' as we went in search of doppelgängers, an expedition that relied heavily on the concept of entropy. We said at the time that entropy counted microstates, but this was a slight oversimplification – it is only true when all microstates are equally likely. It was Gibbs who showed how to generalize to more generic situations. If the first microstate is tagged with probability p_1, the second with probability p_2, the third with p_3, and so on, then a more precise formula for the entropy is given by

$$S = -p_1 \ln p_1 - p_2 \ln p_2 - p_3 \ln p_3 - \ldots$$

The similarity with Shannon's formula is stunning. The only difference, of course, is that Gibbs uses natural logs while Shannon uses logs of base 2. The truth is that this difference is just a matter of convention. Shannon chose base 2 because he wanted to measure information in bits, to compare with a single binary outcome like the toss of a coin. But that is just a choice. You could just as well measure information in nats. One nat is equivalent to around $1 / \ln 2 \approx 1.44$ bits. It compares information to $e \approx 2.72$ possible outcomes, rather than two. For whatever reason, nature prefers to trade in nats rather than bits and, if we switch units accordingly, Shannon's formula coincides *precisely* with that of Gibbs.

Are entropy and information really the same thing? I would say yes. Both of them measure a degree of mystery and uncertainty, although they come at it from slightly different perspectives. We talk about the entropy of a gas or an egg or a triceratops because we can't be sure what state it is really in. There are a bunch of things we don't know or care to know. By any practical definition, the triceratops is the same triceratops even if we flip the spin of one of the electrons deep inside its intestine. Entropy counts all this uncertainty and indifference. But imagine if you weren't indifferent and you decided to measure the spin of that electron and everything else you weren't sure about. You would gather an awful lot of information. How much? Well, that is given by how much uncertainty there was to begin with, and that is just the entropy.

Information is more than just an abstract idea. It is physical. We can even ask how much it weighs. The precise value depends on how the information is stored. For example, the data on your mobile phone is stored by trapping electrons on a memory block. The electrons stored in the trap have a higher energy than those that aren't and because they have more energy they weigh more. That has to be the case because mass and energy are equivalent, as Einstein explained through the poetry of his most famous equation, $E = mc^2$. On average, one bit of data adds about 10^{-26} milligrams in weight. To increase the weight of your mobile by the weight of a speck of dust you would need it to store around 10 trillion gigabytes of data.[3] According to the International Data Corporation, that is about the size of the *global datasphere* making up all the world's collective data.

We have become quite adept at storing information. When eighteenth-century textile worker Basile Bouchon figured out how to control his loom using perforated tape, he was only able to store a handful of bits in a few centimetres of roll. To compete with the 64 gigabytes I have stored on my iPhone, Bouchon would have needed a roll of tape stretching ten times further away than the Moon. Data is being squashed and squeezed into ever-decreasing spaces as technology accelerates to keep pace with demand. Will Apple ever release a phone that can store those 10 trillion gigabytes?

They already have.

My iPhone may use its electron traps to store up to 64 gigabytes of photos, videos and WhatsApp messages, but it stores far more information elsewhere, in the full network of atoms and molecules from which it is built. The trouble is that this additional information isn't particularly useful to us. We can't read it or manipulate it. We can estimate how much of it is there by working out the thermal entropy of the phone. That is around 10 trillion trillion nats, or around 1,000 trillion gigabytes.[4] As you can see, there is a huge amount of data contained in this microscopic structure, but you cannot use it to show Grandma a video of the kids playing with the dog in the back garden. Perhaps one day we will figure out a way to store one bit of data on each of its atoms, or maybe even on each of its quarks and electrons. Only then will the storage capacity of a mobile phone start to become comparable to its thermal entropy. If and when that happens, we can really start to speculate about our ability to store data in ever more confined spaces.

But there will come a time when data gets claustrophobic. Black holes are the problem – they put a cap on how much data you can squeeze into a confined space. The reason is that they also carry entropy. This has to be the case: if they didn't, what would happen if you threw a politician into a black hole? That politician carries plenty of entropy, from the arrangement of atoms and molecules in their feet to the misinformation stored in the neurons of their brain. Once they have disappeared behind the horizon and become one with the black hole, their entropy would be lost. The total entropy would be seen to decrease, in violation of the second law of thermodynamics. To protect the second law, if not the politician, the black hole *must* pick up the entropic tab.

You can get an instinct for how much entropy there is inside a black hole by looking at what happens when they get cannibalistic. If one black hole consumes another, the total horizon area will always increase. This need to ascend in area mirrors the ascent of entropy we saw in thermodynamics. Jacob Bekenstein took the connection seriously and, in 1972, proposed that the entropy of a black hole should be related to the area of its event horizon. But Bekenstein's idea needed proof. It needed calculation. And for that it would need the bravery and brilliance of a young physicist called Stephen Hawking.

We have already seen that Hawking calculated the entropy to be

$$\frac{A_H}{4l_p^2}$$

where A_H is the area of the horizon, and l_p is the Planck length. What is remarkable is how he did this. Up until the mid-1970s, black holes did exactly what they said on the tin – they were black. Or so people thought. Then Hawking did the unthinkable – he challenged that. He took the defining feature of a black hole – the fact that all particles, including light, were gravitationally bound – and showed that it wasn't true. To many, it was moronically oxymoronic. But Hawking wasn't being reckless. He just realized that quantum mechanics was the one way out of nature's Alcatraz.

In quantum theory nothing is as quiet as it looks. As we will see in the chapter '10^{-120}', the silent emptiness of space is really a bubbling soup of *virtual* particles, popping noisily in and out of existence. Virtual particles aren't really particles at all – what they really represent is an identity crisis. When we talk about a *real* particle, we mean a localized ripple in some particular field: a photon is a ripple in the electromagnetic field; a graviton is a ripple in the gravitational field; an electron is a ripple in the 'electron field'. The trouble is that quantum mechanics can blur this description, at least if two fields know how to interact with one another. If a neutron is moving through a gravitational field, it isn't *always* just a ripple in the neutron field – it spends some of its time also rippling the gravitational field. Likewise, ripples in the gravitational field will spend some of their time rippling the neutron field. Let us draw an analogy. Imagine two people from very different backgrounds: one grew up surrounded by socialism, the other in a much more conservative environment. You should think of Lefty as a ripple of the left-wing field and Righty as a ripple of the right. Both are the products of their environments, bred with an assured confidence in their respective ideologies. And then they meet – they interact. They are reasonable people, so they don't just talk, they also listen. As a result, there are moments when they aren't so well-defined. Lefty is still to the left but at times will pause to consider the wider economic impact of his radical ideas. For her part, Righty still likes to think of herself as a conservative, but occasionally she worries about social justice and the problems of inequality. You can think of virtual particles as being a bit like this contamination of ideas. However, this dalliance with other ideologies is only ever a brief consideration. Lefty always winds up true to his socialist ideals and Righty to her conservatism. So it is with virtual particles – you never find one you can hold on to for ever. The rippling into other fields is always temporary.

Hawking was thinking about this sort of contamination in the vicinity of a black hole when he realized something remarkable – what you thought was only temporary can sometimes become permanent. If a pair of virtual particles are produced close to the black hole horizon, it is possible for one to fall in and the other to escape. The escapee, separated for ever from its partner, now becomes a *real* particle, a permanent relic you can actually grab hold of. It behaves just as if it were radiation emitted from the event horizon, drawing energy from the gravitational field, weakening it a little. The result is a stream of radiation – now known as Hawking radiation – and a black hole that is evaporating.

Hawking was able to show that his radiation endowed the black hole with a temperature and, with a bit of thermodynamic gymnastics, derived the entropy formula. Academically, it was an astonishingly brave move, so radical was his proposal at the time. But Hawking's bold genius was rewarded and his ideas are now universally accepted.

Having just announced that black holes aren't actually black – that they give off radiation – Hawking immediately dropped another bombshell: quantum mechanics was broken.

Many countries have a written constitution, a basic set of rules laid down by the founding fathers spelling out their vision for their fledgling nation. The same is true for the nation of quantum mechanics. It has its own constitution, a series of fundamental postulates written out by the early quantum pioneers: Bohr, Heisenberg, Born and Dirac, to name just a few. One of these fundamental rules states that nothing is ever lost: what goes in must come out. Hawking realized that a black hole seemed to ignore this: it would start out as a pure quantum state but end its days as radiation, which he described as a mixed state. We came across pure and mixed states in the previous chapter. Pure states tell you everything there is to know about the quantum system, in contrast to mixed states, where some information is missing. The point is that the quantum constitution forbids any loss of purity – you cannot pass from pure to mixed – because information should not just disappear. It should always be there somewhere, even if it is a little hard to find. Black holes seemed to be rebelling against quantum mechanics.

This is known as the *information paradox*. It is one of those puzzles that is so profound its resolution is expected to reveal something really important about the world in which we live. Hawking liked to bet on this sort of thing. In 1997, he and Kip Thorne signed a wager with Caltech physicist John Preskill: Preskill was sure that the information could never be lost, even inside a black hole; Hawking and Thorne thought otherwise. Whoever was proven right would receive an encyclopaedia of their choice. It was a suitable prize, given that the bet really rested on whether or not you could reproduce the information contained in the encyclopaedia, even if someone were careless enough to drop it into a black hole. Seven years later, Hawking proposed a resolution to the paradox and conceded the bet. He sent Preskill a copy of *Total Baseball, The Ultimate Baseball Encyclopedia*, joking that he ought instead to have burned it and sent the ashes. After all, the information should still be there! For his part, Thorne didn't pay up because he wasn't convinced by Hawking's proposal. The truth is it didn't catch on. Even so, there are very good reasons to believe that black holes don't rebel – that information is not lost – and we'll explain why in the next chapter. Quantum mechanics is just too precious to cast aside.

For their size, black holes carry an enormous amount of entropy. That enables them to store huge amounts of information – information we now believe we can access in principle although never in practice. A black hole the size of my iPhone would store a whopping 10^{57} giga-bytes.[5] That dwarfs the 64 gigabytes I have stored in photos and messages or even the 10^{15} gigabytes there is contained in its atomic information. Nothing stores information as efficiently as a black hole.

To see why, suppose you are an intergalactic traveller on a mission to visit an exo-planet known as Kepler-62f, almost a thousand light years from Earth. Kepler-62f is in orbit around Kepler 62, a star that is slightly smaller and cooler than our own Sun, in the constellation of Lyra. There is reason to go there. Kepler-62f has already been identified by SETI as a good place to search for extraterrestrial life. This is an old and rocky planet set in the habitable zone around the star, its surface covered by ocean and with seasons similar to our own. You are travelling there in a spaceship that is not too big – just able to squeeze inside a sphere 3 metres in diameter. The ship is packed full of stuff – food, fuel and, above all, a huge amount of information crammed into the computer systems. The total mass is around a million kilograms. You aren't quite sure how much entropy it contains, but you know it is a lot because of all that information.

As you approach Kepler-62f, you notice something worrying: a gigantic shell has formed around your spaceship. You are suddenly cocooned inside this extraterrestrial sphere. You cannot be sure where it came from, but it doesn't feel like an accident. You are convinced you have been set upon by the inhabitants of the planet and that they have captured you inside their spherical prison. You decide to run some tests. You realize that the cocoon is made up of some seriously dense material, even more dense than a neutron star. This panics you a little. You calculate the total mass of the shell to be just shy of 10^{27} kilograms. Now you are panicking a lot. How is this shell maintaining its shape? Why isn't it breaking up or radiating away its mass? None of it makes sense, but what really bothers you is that the shell appears to be shrinking. You do the maths. The combined mass of you and the shell takes you past a threshold of 10^{27} kilograms. If that shell were to shrink down to a diameter of just 3 metres, within touching distance of your spaceship, you would be cramming too much mass into too small a space. A black hole would inevitably form.

In the end, unfortunately, you died long before the shell shrank within the 3-metre threshold, torn to shreds by the tides of gravity. Afterwards, the inhabitants of Kepler-62f send a probe to examine the black hole that encased your ship. Their goal is to establish how much you knew – how much information did you have on board your ship before you were consumed? They measure the diameter of the event horizon. It is just 3 metres across, and they figure out that the black hole entropy is around 2.7×10^{70} nats. The aliens know that the total entropy cannot decrease over time. Although there might have been a lot of information on board your spaceship before you were captured, they know that it couldn't have been more than the final tally of 2.7×10^{70} nats.

The story is a little fanciful, of course. There is no way the aliens of Kepler-62f could create and control a cocoon of such high density. But it doesn't matter. This is really just a thought experiment and one that was developed by the wonderfully creative American physicist Leonard (Lenny) Susskind. His goal was to show how black holes put a limit on the amount of entropy you can store in a confined space. Take any object – a spaceship, a triceratops, or even just an egg – and place it fully inside the smallest sphere you can manage. Susskind showed that the entropy of the object cannot exceed the entropy of a black hole whose horizon aligns with the sphere. In our fanciful tale, the spaceship could just about fit inside a sphere of 3 metres in diameter. The aliens then showed that its entropy was limited by the entropy of a black hole of exactly that size.[6]

We can apply Susskind's idea to a human head. To put an absolute limit on the amount of information it can store we just need to calculate the entropy of a head-sized black hole. If you ever try to go beyond that limit – if you try to cram too much data into your limited head space – your head is guaranteed to gravitationally collapse. You will become the latest victim of black hole head death.

THINK OF A NUMBER

I don't always think. My wife said I wasn't thinking when I decided to drain the dishwasher using the vacuum cleaner. Yes, I know full well that water and electricity are a dangerous combination. My plan had been to suck the water into the hose and then quickly cut the power. If all went well, I would transfer the water to the sink before it made contact with any of the electrics. Luckily, my wife came home and put a stop to this behaviour before I had the chance to break the hoover and myself. I guess this is why I'm not an experimentalist. I'm OK with a pen and paper and a tricky calculation but, whatever you do, do not let me near an expensive piece of kit. The formidable German Wolfgang Pauli, a quantum pioneer who will star in the second half of this book, had similar issues. It was said that he could break an experiment just by being in the vicinity – so I reckon I'm in decent company.

Occasionally, however, I do think. Normally I think about football or physics or, if I'm feeling particularly reckless, I might even think about numbers. When any of this happens there are certain events that unfold in my brain. What does the brain do when it thinks of a number? What does it need to do to think of really big numbers? And what will happen if it takes on something as large as Graham's number?

Memories, bits of knowledge and perhaps even the last five hundred digits of Graham's number are all stored in the brain through different patterns in a network of neurons. At any given moment, some neurons are at rest, while others are fired. Typically, the brain tries to get away with firing as few neurons as possible. In total, there are about 100 billion neurons in the human brain. Given that each neuron is either on or off, this suggests a storage limit of around 100 billion bits. That is far in excess of what we actually need, unless we decide to take on Graham's number. You might hope to picture its decimal representation in your mind's eye, as long as you could empty your mind of all its inessential information. You could try to forget who your family were or what an egg looked like or how to recognize the sound of a bird singing. Once you were in this state of meditation, you could try packing in Graham's number, digit by digit, using ever more elaborate patterns of neurons. But even if you could manipulate your mind in such a radical way, you would still come up well short. The trouble is that there are far more than 100 billion digits in Graham's decimal representation. You couldn't even picture a Sun tower, never mind Graham's number.

To do better, the brain would have to learn to store information in a more efficient way. We know that nothing is more efficient at storing information than a black hole – could your brain find a way to mimic those storage tricks, whatever they might be? Gia Dvali, a director at the Max Planck Institute for Physics in Munich, has suggested that this might be possible for certain types of neural network. The logic relies on some very exciting ideas about black holes and how they store their information. Remember, this is still something of an open question, so we are talking about research that is right at the cutting edge. To begin with, Dvali and his collaborators reckon that a black hole behaves like a *Bose–Einstein condensate*. This is a very special state of matter in which a very large fraction of the particles exist in the same quantum state, with lowest possible energy. You can form a Bose–Einstein condensate by cooling gas of extremely low density down to the coldest of temperatures, in and around absolute zero, as was done for the first time in 1995, with rubidium atoms. What makes these condensates weird is that they exhibit quantum behaviour even at a macroscopic level. Dvali has a way of thinking about black holes as a condensate of a huge number of gravitons – quantum ripples in the gravitational field – packed in as tightly as possible. Information is then stored in the quantum ripples of the condensate itself. It turns out to be a very efficient way to store data – huge amounts of information at very little energy cost – which is exactly what you expect to get from a black hole. But he has gone further and cooked up a model for a neural network that is able to store information in a very similar way. So what if your brain were capable of storing information using the same sort of neural network?

It still wouldn't be enough to contain Graham's number.

This really boils down to the amount of data you can actually cram into a human head. What is the maximum? To answer this, I decided to look at my own head. which I estimated to have a radius of around 11 centimetres. If we used Hawking's formula, we would find that a black hole of the same radius would carry a huge amount of entropy, equivalent to 10 billion trillion trillion trillion trillion gigabytes of information. This is the maximum amount of information that anyone or anything could ever hope to store in a region of space the size of my head. Contrast it with the Large Hadron Collider, a machine that delights in producing obscene amounts of data. That produces only around 10 million gigabytes of data in an entire year. But 10 billion trillion trillion trillion trillion gigabytes is nowhere near enough to picture the whole of Graham's number. It's not even close.

What about your head? Could you do any better? Every human head is bound by roughly the same limit – around 10 billion trillion trillion trillion trillion gigabytes. Of course, the reality is that your head is never going to get close to storing such a vast number of gigabytes, certainly not while you are still alive. Remember, information weighs, so to get anywhere near that threshold you would have to pack an awful lot of mass into a relatively small head space, more than ten times the mass of the Earth. As you took on more mass and more data, you would encounter tremendous internal pressures and alarmingly high temperatures. Your head would certainly explode, probably more than once. Survival would be completely out of the question.

But we shouldn't let death get in the way of an interesting thought experiment. Imagine if your lifeless body, and what remains of your head, were taken far away by friends, to the unseen depths of interstellar space. Away from prying eyes, they respect your wish to keep chucking in data, more and more of Graham's number, digit by digit. If they could somehow keep enough of that data contained inside your head, eventually it would hit that 10 billion trillion trillion trillion trillion gigabyte threshold. At that point, there wouldn't be a head – there would be a mini black hole. When you cram that much data into such a small volume, a black hole is the *only* physical object capable of storing it.

There wouldn't be a body either. It could never survive intact that close to a head-sized black hole. There isn't a lot that could. You may think that the black hole isn't really *that* big so it shouldn't be that destructive. But remember, inside the space that was once your head is the mass of ten Earths. The gravitational impact of an object like that should not be underestimated. With black holes, it is the little ones you should really worry about. A head-sized black hole would be far more dangerous than Pōwehi, the leviathan we encountered at the end of chapter '1.000000000000000858'. Because it is so small, anything within touching distance of its horizon is far too close to the singularity and is destined to be torn apart by the tides of gravity. It only takes around 10,000 Newtons to tear apart a human body. On the edge of a head-sized black hole, your body would be hit with a tidal force more than a trillion times greater.

Little black holes are terrifyingly real. Of course, the ones you encounter in nature are not the result of some poor wretch being force-fed Graham's number. Nor are they the product of stellar collapse. These diminutive dragons are typically born in the primordial soup of an infant universe. When the universe was a baby it was hot and filled with a bath of radiation. This bath wasn't exactly smooth. Ripples in energy bubbled on top and, in some places, the ripples were so compact they fell victim to gravitational collapse. The resulting black holes were small, far smaller than anything you could create from a star. Some were too small and have long since evaporated away through their Hawking radiation. But anything bigger than a trillionth of a millimetre could still be around today, and that includes black holes the size of your head. There has been plenty of speculation that these primordial objects could be one of the main ingredients of dark matter – the mysterious invisible substance making up most of the matter in the universe. Our own galaxy is shrouded by a vast halo of this stuff, much more abundant than the stars we can actually see. For their part, head-sized black holes could be responsible for up to 10 per cent of it. So, like I said: these things are real. The galaxy may even be full of them.

We have almost finished our experiment. You have finally experienced black hole head death and are now a head-sized black hole, forlornly floating through interstellar space. In truth, you have become an abomination, more likely to be mistaken for dark matter than the human you once were, tearing to shreds anyone who tries to get close to you. And for what? Ten billion trillion trillion trillion trillion gigabytes of data, some of which you are losing to Hawking radiation. And I'm sorry to tell you, but you haven't even scratched the surface of Graham's number.

So, you carry on.

Your friends keep feeding you data. Another of Graham's digits. And another. And another. Your black hole will grow, its event horizon reaching out further and further. It has to grow to acquire more entropy and more information. Eventually you reach the size of Pōwehi. At that point you hold 10^{86} gigabytes and yet still, you haven't scratched the surface of Graham's number. On the upside, you aren't quite as dangerous as you once were. Because you are so big, the tidal forces near your horizon are only small. If a loved one came to kiss you, they wouldn't be torn apart. Admittedly they'd struggle to avoid falling in, but if they could somehow get away, there is some hope that they would do so without being ripped to shreds. It's a small mercy, but a mercy, nonetheless.

You carry on.

More digits. More data. Eventually, the event horizon of the black hole will extend across billions of light years, filling most of the observable universe. At this stage, it begins to sense something new and unexpected: your *de Sitter horizon*. We should take a moment to explain what this is, as it's kind of a big deal.

We live in an unusual universe. Back in 1998, two teams of astronomers, led by Adam Riess and Saul Perlmutter, noticed something strange. They had been watching star death, collecting light from faraway stars as they performed their final hurrah – as they turned supernova. But the light was dimmer than expected, as if the stars were more distant than we had previously thought. It pointed to a quickening. These stars were further away because space was expanding faster and faster. It was *accelerating*. We didn't anticipate this because gravity attracts. You would expect it to slow down the cosmic expansion, its relentless embrace pulling spacetime together. But no, something is pushing the universe apart.

What would do such a thing? We call it dark energy, but that is just a name, a label we give to some unknown perpetrator like Jack the Ripper or The Bogeyman. There are many who say that dark energy is tied to the vacuum of space. This makes sense in our quantum universe, where the vacuum is a bubbling broth of virtual particles, filling the desert between stars and galaxies. You shouldn't think of this broth as something you can hold or capture – you can never hold a virtual particle – but you can feel its effects as it contaminates the fields of gravity, pushing the universe out, forcing it to grow at an ever-increasing rate. A universe accelerated by the broth of its vacuum is known as *de Sitter* space, named after the Dutch physicist who first asked what it would be like to live there.

Riess and Perlmutter's supernovae seem to suggest that this is where we are headed, towards de Sitter, stars and galaxies spreading further and further apart, leaving nothing but vacuum and the accelerating broth. If this is the case, and most physicists now believe that it is, then each and every one of us is surrounded by a vast cosmological shroud almost a trillion trillion kilometres in diameter. This shroud is a horizon of sorts, although it is very different to the event horizon marking the edge of a black hole. It is known as a de Sitter horizon, marking the boundary of what we can ever hope to see, even if we lived for ever. You might think it strange that such a boundary could exist. After all, if you wait long enough, surely there will be enough time to receive light from even the most distant of stars and galaxies. But this is not so. As the acceleration kicks in, these faraway stars quicken their escape. The space between you and them is growing too quickly for light to keep pace. Even in eternity, you can never see beyond your de Sitter horizon. The light from these faraway realms can never hope to reach you.

The word *horizon* is used whenever there is a limit to what you can see. However, it is important to realize that a de Sitter horizon has more in common with an ocean horizon than it does with the event horizon of a black hole. It is not the entrance to a prison or a cloak for a dreaded singularity. And nor does it have an absolute location. Like an ocean horizon, it is a relative phenomenon, a personal thing. Every single individual can describe their *own* de Sitter horizon, a vast cosmological sphere with themselves at the centre. You have your own de Sitter horizon – your own boundary between what you can see and what you can't – and it is in a different place to mine, or to the horizon of an alien on the edge of Andromeda. If you wanted to, you could cross the alien's horizon and they could cross yours, just as readily as a distant ship may disappear below the horizon of another out on the open ocean.

Let's finish our experiment. As you gather more and more of Graham's data, your de Sitter horizon looms large. Your black hole event horizon continues to grow, reaching out further and further, until eventually it kisses your de Sitter horizon. This is known as the *Nariai limit*. You cannot grow your black hole any larger. Your friends could try to pile in more data, to push you beyond your *own* cosmological shroud, but things would go bad. The equations suggest that nature would resist, forcing the universe to collapse into a crunch. And, despite everything you have endured, you would still be nowhere near Graham's number.

In the end, if you really want to capture all the data in Graham's number, you are going to need a bigger universe. If it has a de Sitter horizon, it should be at least Graham's number in size, in metres or miles or whatever units you choose to use. That is not where we live – our de Sitter horizons are tiny in comparison – but a universe like this could exist in principle. String theory predicts a multiverse, a great many universes with different size, shape and dimension. If the multiverse has giants whose cosmological shrouds are unimaginably large, there might just be room for Graham and his gargantuan number.

TREE(3)

THE GAME OF TREES

At forty-seven games all, final set, the scoreboard failed on Wimbledon's court number 18. It was the summer of 2010 and French qualifier Nicolas Mahut and his American opponent John Isner were in the middle of making history. This was already the longest tennis match on record, and it was still far from over. The scoreboard failed because matches like this were not supposed to happen. The engineers who had programmed it had not expected it to take on so much data over so many games. With the scoreboard displaying a blank, the umpire continued to keep score and as darkness descended at the end of the second day's play, the match was still locked at fifty-nine games all. The programmers were able to fix the scoreboard overnight but warned: 'As long as they don't play more than twenty-five more games, we're okay. If they play more, it will break.' They got lucky. In the twentieth game of the third day, Isner struck a stunning backhand pass straight down the line to break Mahut's serve. A war of attrition was finally over. Isner had won: 6–4, 3–6, 6–7, 7–6, 70–68. An unremarkable first-round match had developed into something astonishing. The two men had battled on court for over eleven hours, exhausted but refusing to give in to their exhaustion. They had each served over a hundred aces. To the crowd on court number 18 and the millions watching at home, it was a match that had threatened an eternity.

Wimbledon will never see anything like this again. In 2019, nine years after Mahut and Isner's epic encounter, the All England Club decided to change the rules. They were concerned about scheduling and the impact of marathon duels on the physical strength of the players. They declared that the final set should go immediately to a tiebreaker as soon as the score reaches twelve all. The threat of an eternal match was diminished, but it was not gone. There is no limit to the length of a tiebreaker or indeed an individual game. Tennis still has the capacity to last for ever.

The same is true of Monopoly. I'm sure you have been there, several hours into a game, wondering if it will ever finish, hoping to land on a hotel in Mayfair just so that it will all be over. The threat of eternity will always be present, unless you stick to games that are guaranteed to end after a finite number of steps, like noughts and crosses. Chess is another finite game: assuming we invoke the mandatory seventy-five-move rule, a game of chess is guaranteed to end in less than 8,849 moves. So if the aim is to remain finite, what should you do if someone suggests playing the Game of Trees? Does that threaten you with eternity?

This was the question posed by the great traveller Paul Erdös as he gossiped his way through the mathematical world in the late 1950s. Erdös often spoke of a young Hungarian mathematician he had met in Budapest when both were still in their teens. His name was Endre Weiszfeld or, as he would later become, Andrew Vázsonyi. He had changed his name in response to rising discrimination against the Jews in the 1930s and eventually he would flee to America. According to Erdös, Vázsonyi conjectured that the Game of Trees would always be finite. But he had never proved his claim and now he was dead. In truth, Vázsonyi was alive and well, at least when Erdös was telling his story. But 'dead' was Erdös-speak for having left academia to take up a well-paid job as an aircraft engineer. Even so, the conjecture remained unproven. In the corridors of Princeton, a talented young student was listening with interest to Erdös' gossip. His name was Joseph Kruskal.

In the spring of 1960, after he had just completed his doctorate, Kruskal proved that the Game of Trees is guaranteed to end after a finite number of moves. But be warned: although the game is finite, it could easily last beyond the lifetime of a human, a planet or even a galaxy. You could still be playing up to and beyond the death of the universe.

Let's play.

The idea is to build a forest of trees from a particular choice of seeds.

Here is a typical tree:

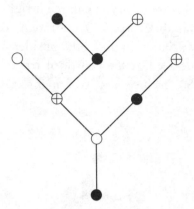

As you can see, trees are just blobs connected by lines. The blobs are the seeds, and the lines are the branches. There are three different types of seed in this particular example: the black, the white and the cross. The rules of the game are as follows: as you build the forest, the first tree must have at most one seed, the second tree must have at most two seeds, and so on. The forest dies whenever you build a tree that contains one of the older trees. There is a precise mathematical meaning to 'contains one of the older trees', but perhaps it is enough to think of an apple tree. Apple trees can stand alone, or they can grow off other trees. Perhaps somewhere in the forest you see a particular apple tree, and then further in you see a large pine with an exact replica of that old apple tree shooting out of its trunk. That's not allowed in the Game of Trees.

To be more precise about this, let's compare some trees and ask if one 'contains' another. For example, consider the following trees, all of which are different:

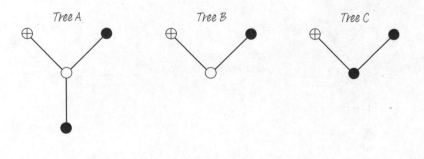

Does tree A contain tree B? The answer is rather obvious. Of course tree A contains tree B – you can see it right there in its uppermost branches. What about tree C? Does tree A contain that? At first glance you might say that it doesn't, but consider what happens when you cover the white seed at the centre of tree A. What remains is essentially tree C. So, in this sense, you might argue that tree A *does* contain tree C.

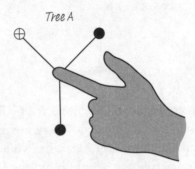

Tree A

To settle the debate, we need to take a closer look at the rule book. For one tree to contain another, we must be able to match up the relevant seeds, just as we did in the example above, covering up the white seed of tree A. But this, by itself, is not enough. The matching seeds must also agree on their *nearest common ancestor*. For any two seeds in the upper branches of a tree, you can work out their nearest common ancestor by tracing their lineage back towards the root and finding the seed where the two lines come together. Imagine yourself and your first cousin as the seeds. If you both traced your ancestral lines, you would come together at your grandparents.

Tree A

Tree C

Nearest common ancestors for
⊕ and ●

Take a look at the black seed and the cross in the upper branches of both tree A and tree C. Tracing the lineage in each case, we see that their nearest common ancestor in tree A is a white seed, while in tree C it is a black seed. Thus, we have a disagreement. It is in this more subtle sense that we say that tree A *does not* contain tree C.

Let's flesh this out some more with a final example. We have two more trees:

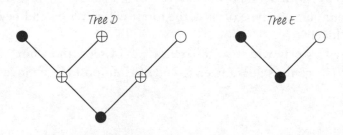

Does tree D contain tree E? The first thing to check is: can we match up the seeds? By covering up all the cross seeds in tree D, we see that we can. Now we need to ask about ancestry. Consider the white seed and the black seed in the upper branches of both trees. Tracing the lineage, we see that in both cases, the nearest common ancestor is the black seed at the root. This ticks all the boxes. Tree D definitely contains tree E.

Now that we understand the rules, we are ready to play. Let's play a game where we are only allowed to use the black seed. I will go first. Remember, this is the first tree, so it can have at most one seed. I'll just scribble it down:

●

Now it's your go. You are immediately in trouble. As this is the second tree in the forest, you could include up to two seeds, but that doesn't help. There are only two possible trees you could write down – another single-seeded tree or a double-seeded tree:

The problem for you is that my tree is obviously contained in *both* of these. If you plant either of them, the forest will die. There is no way to avoid this – the game is over after a single move. When you play with just one type of seed, the forest will never extend beyond a single lonely tree.

Now let's play with two different types of seed: the white and the black. The game is guaranteed to end after at most three moves:

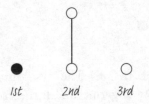

Whatever tree you plant next is destined to destroy the forest. I'm guessing you are not too impressed. Who wants to play a game that will definitely end after three measly moves?

Wait.

The time has come to play with three seeds: the black, the white and the cross. Let's play a handful of moves:

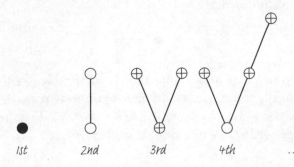

This is good – the forest is still alive. But how much further can we go? We know the game is guaranteed to end at some point – Kruskal told us that – but *when* will it end? After a hundred moves? After a googolplex? After Graham's number?

We aren't even close.

In this book we have already encountered tales of numerical gargantua, of numbers of unfathomable size. But these colossi are nothing in comparison to our next leviathan. This is the number known as TREE(3), the gigantic limit of the three-seeded game. It is part of an utterly bizarre sequence – the TREE sequence. If you play the Game of Trees with *n* different seeds, the game will be done after TREE(n) moves. Look at how it starts off gently:

TREE(1) = 1 (because a single-seeded game will last only up to one move)
TREE(2) = 3 (because the two-seeded game will last only up to three moves)

and then boom!

TREE(3) = a number large enough to engulf a googolplex and Graham's number.

Everything you think you know is dwarfed into nothing. You could go on to even bigger numbers: TREE(4), played with four seeds, TREE(5), played with five, and so on. But TREE(3) is already enough. Breathtaking, unimaginable and absurd.

Vázsonyi's original conjecture and Kruskal's subsequent proof told us that the Game of Trees would always finish, as long as you played with a finite number of seeds. But it was the American mathematician and philosopher Harvey Friedmann who realized it could still unleash some tremendously large numbers. Friedmann had a talent for logic that was already evident at a very early age. When he was just four or five years old, he found a dictionary and asked his mother what it was. 'It's for finding out what words mean,' she told him. A few days later, he challenged her. Dictionaries were useless, he said, because they went round in circles. If you looked up the word 'large', you were taken to 'big', then to 'great' and then back to 'large' again. How could you ever find out what something actually meant? A decade or so later, his precocious talents would gain him a place in the *Guinness Book of World Records* as the youngest university professor, joining the philosophy faculty at Stanford University at just eighteen years old.

Friedmann saw that TREE(3) was spectacularly big. He couldn't pin it down exactly, but he was able to show it was bigger – much bigger – than any other number you will find in this book. He cooked up an estimate – an *under*estimate – in terms of some enormous numbers known as Ackerman numbers. To get a feel for their size, we need to revisit Graham's ladder. You may remember that the first rung was already monstrous, $g_1 = 3 \uparrow\uparrow\uparrow\uparrow 3$, and from then on, things got quickly out of hand. The second rung was built with g_1 arrows, $g_2 = 3 \uparrow^{g_1} 3$, the third with g_2 arrows, $g_3 = 3 \uparrow^{g_2} 3$, and so on, until we reached the sixty-fourth rung and Graham's number. But suppose you kept on climbing, to the sixty-fifth rung, with a Graham's number of arrows, to the sixty-sixth, to the sixty-seventh, to the googolth rung on the ladder. In fact, suppose you didn't rest until you had climbed this many rungs:

$$2 \uparrow^{187,195} 187,196$$

There are 187,195 of Knuth's arrows here. This is a tremendously big number, but all it is doing is *counting the rungs* on Graham's ladder! Sixty-four steps up Graham's ladder took you to Graham's number. Can you even begin to comprehend where $2 \uparrow^{187,195} 187,196$ steps would take you? This true giant is similar to Friedman's estimate of TREE(3), but be under no illusion, it's very much an underestimate. In reality, TREE(3) lies far above, a leviathan among leviathans, dominating everything else we have encountered on our journey through large numbers.

There isn't really an intuitive way to appreciate why TREE(3) is so tremendously large. We can get a hint by looking at the games we played earlier. For the two-seeded game, we are forced to play with the white seed from the second round onwards. With only one colour left to use, we run a much bigger risk of finding one tree inside another, and the game is destined to end quickly. However, for the three-seeded game, by the time we have reached the second round we still have two types of seed to play with. That makes a big difference because we can play with combinations, opening up more and more channels for new and exotic patterns of tree. Eventually we will run out of steam, but not for a very, very long time.

Trees like this matter. They crop up whenever there is branching, from decision algorithms in computer science to the tree of life in evolutionary biology. So-called phylogenetic trees are used by epidemiologists to analyse the evolution of viruses and antibodies. They have also been applied to other evolving systems, such as cancer genomes. But Friedman's interest in trees ran deeper than any of this. He was looking for the *unprovable* truth, that which is true but may never be proven, at least not within its own mathematical framework. This has nothing to do with a lack of skill or talent among mathematicians. These are fundamental truths that are guaranteed to remain unproven *for ever*, whatever the strength of counsel. As we will see, the Game of Trees is a game in this mathematical courtroom – a game of unprovable truths.

The unprovable truth strikes at the very foundations of mathematics. Mathematics has grown from a basic set of rules and principles. For example, from the notion of succession – that you can always increase a number by one – you can build up the idea of addition. All you have to do is repeated succession, increasing by one again and again. From there you can develop multiplication, exponentiation, the concept of prime numbers and all the theorems associated with prime numbers. Mathematics is a man-made system that rules itself. It sets up its own foundations, its basic building blocks, and from there we construct the towns and cities of a mathematical universe. These building blocks are known as axioms. The more axioms you start with, the richer and more complex the mathematical universe you create. It makes intuitive sense. If I have only yellow bricks, then every building in the metropolis will be yellow. But if I have yellow and red bricks, I can create more exciting patterns. There can still be yellow buildings, of course, but I can also create buildings adorned with elaborate mosaics of yellow and red. In the chapter 'Infinity', we will explore another example – the boundary between finite and *trans*finite mathematics. With finite bricks you build finite buildings. To take mathematics beyond infinity, it turns out you need a new type of brick known as the axiom of infinity.

Interest in the axioms of mathematics first took hold at the beginning of the twentieth century. Many of the world's leading mathematicians began to believe in a theory of *all* mathematics. All we had to do was find the full set of axioms and from there everything would follow. With these axioms, all that was true could be proven to be true, in principle at least. Mathematics could be shown to be complete and free of contradictions. This mathematical faith was surely inspired by an appreciation of its power and beauty. Mathematics was conquering the universe. Only a heretic would say that it was broken – that it was incomplete.

That heretic was Kurt Gödel, the brilliant Czech philosopher and logician who many consider the heir to Aristotle. In December 1931, with the world gripped by the Great Depression, Gödel proved the existence of the unprovable truth, that mathematics could never be complete. Whatever axioms you chose – whatever your mathematical framework – there would always be true statements that could never be proven. Of course, you could always consider a larger framework, adding a new axiom to help you with your proofs. But then you would run into trouble with new mathematical statements. Axioms and proof can never keep pace with the truth.

Let's return to our metropolis. The city has access only to yellow and red bricks, so you are not surprised to find it dominated by simple buildings built from these two colours. These buildings are like the provable theorems of mathematics. Given enough time and effort, the city's engineers will be able to tell you how they came to be built. However, somewhere in some dark corner of the city, there is sure to be a strange and mystical building. An unprovable. No engineer will ever be able to tell you how *it* was built, at least from the raw materials the city has available. And yet there it is, proud and unmistakable, a looming reminder of Gödel's genius.

As a precursor to the methods behind Gödel's proof, I want to convince you that all numbers are interesting. For suppose it is not the case – that *un*interesting numbers do actually exist. If a number is genuinely uninteresting, it's unlikely to have a Wikipedia page as there would be nothing to write about. But among these uninteresting numbers, there must be a smallest one. For the sake of argument, let's imagine it is 49,732. Well, now I want to write a Wikipedia page about 49,732 so the world will know this interesting fact – that it is the smallest uninteresting number. We've run into a contradiction. It must be that all numbers are interesting.

Gödel's incompleteness proof follows a similar spirit, although it is far more rigorous. Key to his approach was a systematic code, a way for mathematics to reference itself and ask itself questions. Every axiom, every mathematical statement, true or false, had its own code number. You may imagine a particular number associated with a particular statement, analogous to an ASCII code. For example, one number would correspond to the statement 'The square root of two is irrational' whereas another would correspond to '1 + 1 = 3'. The truth or lie of the mathematical statement could then be linked to a property of the corresponding number. For example, you might say that even numbers correspond to true statements whereas odd numbers correspond to false. Of course, in reality it was much more complicated than that, but the spirit is correct. Armed with the rigour of his new coding system, Gödel considered the following statement:

'This statement cannot be proven from the axioms.'

Now let's step outside the system and *assume* that mathematics is free of contradiction. That means Gödel's statement must be true or false. It cannot be both. Assume it is false. This means that the statement *can* be proven from the axioms and you have a contradiction. But that isn't allowed, so the statement must be true. We seem to have found a mathematically true statement that cannot be proven from the axioms – we have uncovered an *unprovable* truth, the mystical building in our mathematical metropolis.

Mathematics can never be complete.

Gödel's theorem made him famous. It appealed to a spiritual ideology, the idea that a mathematical universe was never enough. Despite his success, Gödel's life was troubled by depression and, as time wore on, he grew increasingly paranoid. He became convinced he would be poisoned and only ate food tested and prepared by his wife, Adele. When she took ill in 1977 and was hospitalized, Gödel refused to eat. He eventually died of malnutrition, on 14 January 1978.

Mathematicians wanted to find more interesting examples of the unprovable truth, beyond the contrived example Gödel had uncovered. They had a vested interest. Imagine you are trying to prove (or disprove) a famous mathematical theorem. It could be the Riemann hypothesis or Goldbach's conjecture or one of the many other unsolved problems in mathematics. If you are young enough, the proof will earn you a Fields Medal, so you work your socks off, day and night. If the only unprovable truths are contrived Gödellian statements, your work has a chance of success. But what if there are more interesting unprovable truths? What if the theorem you are working on is true but unprovable in the mathematical framework we have created? Then you have no chance. You are destined to fail.

In 1977, British mathematician Jeff Paris and his American collaborator Leo Harrington showed that the mathematician's greatest fear could actually be realized. By working in a stripped-back version of mathematics known as Peano arithmetic, they were able to express a true statement about Ramsey Theory that couldn't be proven within that particular set-up. So in other words, Peano arithmetic allowed them to conceive of the theorem and state it explicitly, but it would never let them prove it. To do that, one would have to step outside, into a much larger mathematical framework with more axioms. Paris and Harrington's unprovable truth served as a warning to mathematicians everywhere.

Harvey Friedmann was also on the hunt for unprovable truths. His mission was to unpick the theorems of mathematics. He wanted to reverse mathematics, to understand which axioms were required for which theorems. Imagine yourself walking through the city and seeing a yellow house. You ask yourself, what did I really need to build this house? Of course, you would need only yellow bricks. Yellow and red would be overkill. This was the logic Friedmann was applying to mathematics.

Friedman's quest led him to the Game of Trees and the unprovable truths that lurk within. To see them you must first play the game in a finite world – the world of finite mathematics, a mathematical framework built exclusively from finite bricks. Of course, there are many *provable* truths in this particular world. For example, it is easy to prove that TREE(1) and TREE(2) are finite. All we have to do is play all possible games and see how quickly they end. We could also prove that TREE(3) is finite in exactly the same way, at least in principle. I know I said that a three-seeded game could take us beyond the death of the universe, but right now we are just doing maths, not physics (sacrilege!). I will allow us to imagine a future that is long enough for us to play and play for as long as we need. Just by playing a fantastically finite number of fantastically finite games, we can also show that TREE(4) is finite, as well as TREE(5), TREE(6), and so on.

Suppose we stay in this finite world, can we prove that TREE(n) is finite for *all values* of n? Naively, you may think that we can, given all that we have just said. However, this is a stronger statement than saying that TREE(n) is finite for any *particular* value of n, such as three or four or a googol. And yet we know from Kruskal that the stronger statement is also a *true* statement. So we ask again, can we prove it in a finite world, just as one can prove that TREE(3) or TREE(4) is finite? The answer is no. In his proof, Kruskal went *trans*finite and Friedmann realized there was no alternative. So here you have it in the palm of your hand:

'TREE(n) is finite for all values of n.'

An unprovable truth in a finite world.

THE COSMIC RESET

Now I want you to play the Game of Trees again, but this time in the physical world. This time the laws of physics will affect you and your game and the unexpected universe that surrounds you. Thanks to the magnitude of TREE(3), the game can stretch ominously into the future, putting you at the mercy of a cosmic reset, a quirk of our idiosyncratic cosmology and its holographic truth. But we are getting ahead of ourselves: there are lots of other interesting things that could happen long before you reach the cosmic reset. Let's see what actually happens.

You set up the game in a park on a beautiful autumnal day, sunlight catching on golden brown, the silent surroundings broken only by the song of an occasional blackbird. And then you begin. The serenity is destroyed by the pace of your play. Frantic, you play as quickly as physics will allow, a new tree drawn every 5×10^{-44} seconds. This is the Planck time, the shortest time imaginable. To imagine shorter times would break the structure of space and time in ways that we don't yet understand, as gravity falls victim to quantum mechanics. After twenty-four hours, you have drawn a trillion trillion trillion trillion trees, but the game is not over. Remember, it has the *potential* to last for up to TREE(3) moves, and you are nowhere near that limit.

You play for a year and still it goes on. You play for a century and still it goes on. I imagine you as forever young, like Peter Pan, unable to age, answering only to physics as opposed to biology. The centuries turn into millennia, the millennia into mega-annums (a million years), and all the while the game goes on. After 110 million years you notice that the Sun is about 1 per cent brighter than it was when you began, and the Earth is getting warmer. The continents embrace as one, finally coming together as a single supercontinent after around 300 million years. After 600 million years the Sun will have become bright enough to destroy the Earth's carbon cycle. Trees and forests can no longer survive, but still you continue to play. As the oxygen levels fall, deadly ultraviolet radiation begins to penetrate the Earth's atmosphere. You take the game indoors as a precaution. After 800 million years, the Sun has destroyed all complex life on Earth, except you, of course, as you continue to survive against the odds. Three hundred million years later, when the Sun is 10 per cent brighter than it is today, the oceans begin to evaporate.

You play on. As the Earth becomes increasingly uninhabitable, Mars offers some sanctuary. In around one and a half billion years' time, conditions there resemble those on Earth during the Ice Age. You decide to move your game. It is a wise move: after four and a half billion years, the Earth, gripped by a runaway greenhouse effect, is as inhospitable as Venus is today. At around the same time, the galaxies collide: Andromeda and our own Milky Way, giving birth to Milkomeda, the new galactic chimera. In the interstellar chaos that follows, the fate of the solar system is uncertain. Some models suggest that it will veer towards the central black hole, before being spat from the throat of the galaxy like an interstellar ball of phlegm. Not that this will matter much to you. On your new Martian home, warmed by the heat of the brightening Sun, you continue to play.

A billion more years and the Sun runs out of hydrogen in its core. It triggers a metamorphosis, as the Sun begins to grow into a red giant. Over the next 2 billion years, the Sun will expand, consuming Mercury and Venus, perhaps even Earth. Mars gets too hot and you move the game to the moons of Saturn. But the warmth doesn't last. After playing the game for around 8 billion years, the outer layers of the giant drift away and the Sun becomes a white dwarf. It is feeble, half its current mass and barely bigger than the Earth, unable to warm any of the planets that still survive. Of course, it is fanciful to think that you could really survive such epic change and epic times, but if you did, the game would not necessarily be over. The limit of TREE(3) is just too big.

After a quadrillion years, the Sun will stop shining. Perhaps it will wander lonely through the emptiness of space with its planets in tow. Perhaps it will encounter a black hole. We do not know. What really happens in the latter days of the universe depends on dark energy, the mysterious substance that is dominating the cosmological evolution we see today. Right now, we know that dark energy is forcing space to expand faster and faster.

In the last chapter, we said that many physicists believe dark energy to be tied to the vacuum of space. In a quantum universe, this vacuum is expected to be busy, a bubbling broth of virtual particles spreading their energy evenly, right across the desert between stars and galaxies. If this is really the origin of dark energy, then ours is a cold and gentle future, at least for a while. The universe will continue to grow, expanding at an accelerated rate. In around 10^{40} years, most of the matter we see today will have been eaten, consumed by an army of supermassive black holes patrolling the universe. These black holes enjoy a spectacular era of cosmic dominance, reaching all the way to the googolannum, a googol years from now, and then they will just die. They will decay, exactly as Hawking predicted, spreading his radiation throughout an otherwise empty universe.

The radiated photons and subatomic particles spread further and further apart as the expansion continues. Eventually, the emptiness of space is all that is left, but remember, it is an emptiness filled with the bubbling broth of dark energy. You are now in de Sitter space, frozen throughout at a temperature just above absolute zero. This is the cosmic equivalent of dying in your sleep, free of all drama, save for the occasional tickle of thermal fluctuations. And yet, if anyone were still able to play, the Game of Trees could still be going.

But what if dark energy is *not* the bubbling broth of the vacuum? What if our destiny is different to de Sitter space? Then the death of the universe can be much more violent. For example, if dark energy were to one day disappear, in a billion years or more the universe might stop expanding. In fact, it could even contract, collapsing back into itself, squeezing the energy of the universe closer together, ending in an apocalyptic crunch: the Big Crunch. The most frightening thing about the Big Crunch is the speed of contraction. Generically, it is rapid, far quicker than the corresponding expansion. Like a rollercoaster, the universe climbs slowly to the zenith before tumbling down at breakneck speed.

Another possibility is that dark energy grows. It gets bigger, not just accelerating the expansion but accelerating the acceleration. This is the Big Rip, a universe torn apart. As the universe rips, the expansion of space is so violent that planets are torn away from their stars, like a child that is taken from its mother. And yet, there is no let-up. In time, the cosmic expansion rips apart the atom, the nucleus and everything beyond.

As the universe enters its death phase, whatever that may be, what will become of the Game of Trees? For a crunch or a rip, the death is too violent, and the game is cut short. But right now, most scientists would predict a gentler future. From supernovae observations to measurements of the cosmic microwave background radiation, all the evidence points towards a universe dominated by the bubbling quantum vacuum, a frozen de Sitter space. If this is our fate, you can imagine the game for a little while more, played beyond the googolannum, through the gentle death of our universe. The identity of the players will change over aeonian time. It has to; no individual can exist for that long without falling victim to thermal and quantum instabilities. But what of the game itself? Can it go on for as long as we need? Can it reach the limit of TREE(3) trees?

It cannot.

The gentle death is not eternal. After $10^{10^{122}}$ years, just beyond the googolplexannum, the universe will repeat. The universe will repeat.

This is *Poincaré recurrence time* – the time it takes for our corner of the universe to return arbitrarily close to where it is now. It returns to the *same* quantum state, describing the *same* stars, planets, humans, toads and microbial alien life, just as we see them today. The recurrence awaits because you are surrounded by a gigantic sphere of great importance, and within that sphere, the universe can arrange itself only in so many ways – it has only a finite number of outfits. I will explain why this is in a moment, but first you should imagine the universe trying on the outfits. As it steps through time, it takes on a variety of different looks, from how it looked when an asteroid struck the Yucatán to how it looks now, or how it will look when Justin Bieber is elected president. It will try on every outfit for a second time, and then a third, and a fourth, and so on, forever revisiting the glories and failures of its past. For our corner of the universe, the time for recurrence is long, unimaginably so, and yet the Game of Trees is considerably longer. Even with the gentlest of futures, our universe refuses TREE(3). It resets itself, again and again and again, long before the game could ever reach its limit.

Poincaré recurrence, named after the French mathematician Henri Poincaré, is a feature of any finite system, be it our universe, a box full of nitrogen gas, or even a pack of playing cards. As you move through the system, you explore every possibility until, eventually, you return to where you began. And then you go again. There are almost 10^{68} ways to arrange a pack of fifty-two cards. When you first open them, they are carefully presented, grouped according to their suit and ascending in order. And then you shuffle the pack, spoiling the elegant arrangement and replacing it with something new. You shuffle again, rearranging the cards once more. If you shuffled and shuffled and shuffled, for as long as a googolannum, there is no doubt you would see *some* arrangements repeated. But Poincaré proved something stronger. If the shuffles were genuinely random, then at some point the cards would return to the ordered arrangement in which they were bought. That is Poincaré recurrence.

What about the box of nitrogen gas? Suppose you begin with all the molecules squeezed into the top-right-hand corner. As time goes by, you watch them spread out. They dance and collide, exploring an enormous variety of looks, but one day they return. They find themselves gathered in the top-right-hand corner, just as they were in the beginning. Our universe is no different. If it only has a finite number of ways to arrange itself, then by the rules of Poincaré, it will always return to where it is now. It will repeat.

I mentioned something about a gigantic sphere that surrounds you. This is an artefact of our cold and empty future, a future in de Sitter space, dominated by the energy stored in the broth of its vacuum.

Because of this, each and every one of us is surrounded by a giant cosmological shroud known as the de Sitter horizon. I told you a bit about it in the last chapter, but it's worth a recap. You have your own de Sitter horizon, and I have mine. Yours is a gigantic sphere around 17 billion light years in radius with you at the very centre. It represents the limit of what you can ever see. For example, there may be aliens in some unimaginably distant galaxy arguing over some alien form of Brexit, but you will never see the argument, even if you lived for ever. This is because dark energy is pushing the space between you and them apart, at an ever-increasing rate. Of course, light will reflect off the arguing aliens and some of it may even be headed in your direction, but it will never be able to reach you. The space in between just grows too quickly and the light from the aliens is unable to keep pace.

I also told you that a de Sitter horizon is *not* the same as a black hole event horizon. It is not a boundary of no return or a cloak for a murderous singularity. But despite these important differences, there are some aspects in which the two horizons are quite similar. This idea was developed by Stephen Hawking and his former student Gary Gibbons. They were able to show that quantum radiation is being emitted from your de Sitter horizon, just as it is emitted from the event horizon of a black hole. In our corner of the universe, the temperature of this de Sitter radiation is cold, around 2×10^{-30} kelvin, so you could never realistically hope to detect it, but nevertheless, it is there. As the universe dilutes through the relentless expansion of space, this will be the temperature of the frozen emptiness that is left behind. It is like the Nordic hell of Niflheim but with the slightest touch of warmth, lifted from absolute cold by the tiniest temperature. And remember, whenever there is temperature, there is entropy.

Just as the entropy of a black hole is proportional to the area of its event horizon, so the entropy of de Sitter space is proportional to the area of the de Sitter horizon. The de Sitter horizon that surrounds you is huge, with an area of almost a trillion trillion trillion trillion square kilometres. If we use Hawking's famous formula relating this area to entropy, we end up with more than 30 billion trillion googols of entropy. This helps us count the microstates – the number of outfits that your universe has in its wardrobe. That much entropy corresponds to a cosmic wardrobe with $10^{10^{122}}$ different outfits. Although this is big – bigger even than a Kardashian closet – it is *finite*. If we imagine the universe trying on a new outfit every Planck time, or every second, or even every year, then after around $10^{10^{122}}$ changes it will find itself trying on the same outfit it is wearing today. That is its Poincaré recurrence – a fashion faux pas forced on it by a cosmological shroud and a frozen future.

Although the recurrence in our corner of the universe is likely to be real, given everything we know about dark energy, the time scale for recurrence is so long that nothing could ever see it. There is no being or machine capable of such precise measurement and ancient longevity. The problem is quantum instability. Suppose you had the ultimate piece of kit, capable of measuring the state of the universe with astonishing accuracy. It measures the universe today and records what it sees. In every future moment it measures and compares, but to spot the recurrence it would need to survive for a spectacular amount of time. That is impossible. Quantum instabilities will always overwhelm it, spoiling all its records. The Poincaré recurrence of our universe is out there, but no one will catch it with experiment. In a way, it's Gödel's incompleteness, but with physics rather than maths: an unprovable truth of the physical realm. We could say the same of TREE(3) and the Game of Trees. It exists in principle, but it is so big the laws of our universe will never let it happen.

THE HOLOGRAPHIC TRUTH

Our journey through the land of large numbers is almost at an end. We have ventured into the microworld and the macroworld. We have seen the blurred reality of quantum mechanics that lies within everything there is and reached towards the edge of a black hole, where time is at a standstill, and across a universe whose boundary is still unknown. I hope you are beginning to look upon numbers as a gateway to the most remarkable physics in the universe: from a googol and a googolplex to discovering doppelgängers, from Graham's number to the dangers of black hole head death and from TREE(3) and the Game of Trees to a cosmic reset. These gargantuan numbers have pushed our physical understanding to the very edge of what is currently known.

Perhaps you have noticed a theme. At each and every turn, we have been challenged by entropy – by the limits on the number of micro-states that can describe you, your head or all of the universe you could ever hope to see. And yet, despite all the drama that has followed, there is a single physical principle that underpins everything we have discovered. It is closer to the edge of physics and far more dramatic than everything that has gone before. I think you are ready for it. We'll begin with a horror story.

The alarm screams at you to wake. Without opening your eyes, you reach out to smother it into silence. Instinctively, you climb from the bed and stumble into the shower. Warm water flows over your head, dragging you slowly into consciousness.

And then the terror.

You are trapped in the wall, bound to a two-dimensional prison. And it isn't just you. Everything is bound: the shower, the sink, the bed you were sleeping in. A feeling of panic swells inside. You rush back to your room, dress quickly and scamper down the stairs. The sensation is strange. It is as if you are moving through the world you once knew – the world of three dimensions – but now you know that it is really a lie. You have woken into a nightmare. You must escape, so you open the front door.

But the terror will only increase.

The rest of the world is just as trapped as you are, and yet no one seems to know it. A well-dressed woman cycling by. A man who looks harassed, scruffy and late. A bus full of schoolchildren chatting with excitement. All of them flattened and none of them aware of it. You rush towards the woman, but she cycles away quickly, glancing back with a fearful look. Then you fall to your knees. As the terror of your epiphany begins to overwhelm you, you unleash a primeval cry. This is it. This is the reality. You are nothing but a hologram.

This is your story: a physicist awoken, recognizing the holographic reality of the universe. It is where this book has been leading you – the realization that gravity and the three dimensions of space are some-thing of an illusion. You can just as readily imagine yourself in a holographic world, trapped on the boundary of the space that we normally perceive.

I should probably explain.

The holographic revelations began with Bekenstein and Hawking. They figured out that black holes carry entropy just like you and I, or an egg or a triceratops. As usual, that entropy counts up all the possible microstates that could describe the same black hole. It also measures the hidden information. You may remember the black hole at the bottom of your garden, discussed in the chapter 'A Googol.' It changed its mass by that of an elephant, but we couldn't be sure if it had swallowed an elephant or an encyclopaedia with elephantine mass. That meant you could imagine different microstates describing the same macroscopic object. In other words, the black hole had to have entropy.

But Bekenstein and Hawking went further. They realized that the entropy of a black hole grows with the area of its event horizon. You can think of this as the area of its boundary. This is known as the area law for black holes, and it is not what we had come to expect. You see, you and I don't follow an area law, and nor does an egg or a dinosaur. In fact, the entropy of ordinary things like humans and eggs grows with volume, as opposed to surface area. This makes intuitive sense, and we can even use your head as an example. If you wanted to grow its data capacity – or, more precisely, if you wanted it to be able to store more entropy at the same temperature – you would need more neurons. And for that you would need a bigger brain with a bigger volume – not just a bigger skull.

But why do black holes behave differently to the rest of us? Why does their entropy grow with surface area and not volume? The thing that sets you and the egg apart from a black hole is the extent to which you feel gravity's crushing embrace. Black holes are deeply gravitational – it is gravity that binds them and, without it, they wouldn't exist. When gravity becomes that important, the rules for storing entropy are different to what we are normally used to, and the underlying reason will challenge your concept of reality.

In the early 1990s, Dutch Nobel Laureate Gerard 't Hooft and Stanford physicist Lenny Susskind, who we encountered in the previous chapter, began to appreciate what Bekenstein and Hawking's work really meant. As we have already seen, they realized that black holes were at the top of the entropic food chain, putting a limit on the amount of information you can squeeze into any volume of space. The limit is reached when the space is filled with the largest possible black hole and, because of the area law, the limiting entropy is given by the surface area of its boundary as opposed to the volume of its interior. But their great epiphany was this: if the maximum entropy is given by the surface area of the boundary, we should imagine all the information as being stored on the boundary. In other words, if I want to describe the physics inside some volume of three-dimensional space, I might as well encode everything on the boundary of that space – on the two-dimensional surface that surrounds it.

Let's consider this for a moment. 't Hooft and Susskind were telling us that you can find all the information you will ever need on the surface that wraps around the space you are interested in. It's a bit like saying that the true content of any package can always be found in the wrapping. You can imagine such a package delivered to your front door, perhaps even by 't Hooft himself. When you tear it open, you discover a book: *Fantastic Numbers and Where to Find Them*. You glance at the table of contents: why does Graham have a number? What on Earth is TREE(3)? Putting the book down, you gather up the wrapping and throw it into the recycling. But then you notice something: the wrapping isn't blank but is covered in tiny little letters. In fact, if you are not mistaken, you would say that the words are the same as those found in *Fantastic Numbers*. 't Hooft has delivered you a package with all the information stored in the wrapping – on the boundary of the space that it occupied.

Let's be a little more precise with another analogy. Imagine you are given a box of Lego for Christmas, only it isn't just any Lego, but Planckian Lego. There are a huge number of black and white bricks, each of which is incredibly tiny, with sides just a Planck length across (around 1.6×10^{-35} metres). The box comes with a set of instructions on how to build a Lego universe. You start to build and soon enough you have built a universe like the one shown in the figure.

It is just a small one, a random pattern of black and white, arranged in a cube with eight bricks along each side. According to 't Hooft and Susskind's conjecture, we should be able to encode everything we need to know about this universe on its boundary. The boundary is made up of 6 faces, each containing 64 squares, corresponding to 384 squares in total. As there are two possible colours, that should allow us to code up to 2^{384} different patterns. But now we have a puzzle. When you also

A Lego universe.

consider the interior, there are $8 \times 8 \times 8 = 512$ bricks in total, so you could imagine up to 2^{512} possible arrangements. How can 2^{384} patterns encode 2^{512} possibilities? The truth is they can't. If 't Hooft and Susskind are right, there must be patterns in the interior that aren't allowed to happen, that are never allowed to exist, not even in principle. What is it that stops them from happening? Who is the inhibitor? It can only be gravity.

Remember, it was gravity that broke the entropic traditions. It was gravity and black holes and the unexpected area law that led 't Hooft and Susskind to believe that all the information could be stored on the boundary. And so it must be gravity that prevents you from packing one bit of information on to every Planckian brick. In the end, we have two equivalent descriptions of our Lego universe: the interior with its arrangements inhibited by gravity; and the boundary with *all* its patterns and no inhibitions and therefore no gravity. It's just two different ways of describing the same thing. When an Englishman sees a plate of meatballs, he calls it a plate of meatballs, but a Spaniard would call it *albóndigas*. Both men are describing the same thing but in a different language. And so it is with our physical universe. You can describe it with a theory of three dimensions of space and the force of gravity or you can use a different theory, pinned to the two-dimensional boundary of space and without any gravitational force. As soon as we run with this boundary picture, we begin to think of the highest spatial dimension as an illusion. You don't really need it because the boundary theory captures everything. In a way, it *is* everything.

I realize you might find this upsetting. How can you experience three dimensions of space if there is a perfectly good physical description in which there are really only two? It is all to do with how you decode the information. In fact, it is closely related to the way in which you decode a hologram. So how does that work? Well, suppose you wanted to make a simple hologram of a teddy bear. First you need a laser beam, firing pure light of a single colour. This is then split into two daughter beams: one beam is shone on to the teddy and scattered, while the other is bounced off a mirror. The two beams come back together on a high-resolution photographic plate. Because one beam was disturbed by the teddy and the other not, the peaks and troughs of the two waves do not necessarily arrive in perfect unison. In fact, any mismatch is recorded on the plate through an *interference pattern* of light and dark bands.

We encountered similar ideas in the Chapter 'A Googolplex' when we discussed Young's double-slit experiment. The details here are different, but the key principle is the same: if two peaks arrive in unison, you get constructive interference and a lighter band; but if a peak arrives with a trough, you get destructive interference and a darker band. You can now think of this image of light and dark bands of varying intensity as a two-dimensional code for the three-dimensional object, but you still need to do a little work to decode it. If you just stared at the interference pattern on the photographic plate, you wouldn't really see

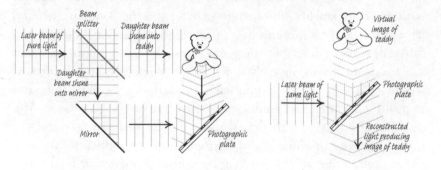

How to create and decode a hologram.

anything interesting. To bring it alive, you shine through another beam of the same light, turning the two-dimensional information back into a three-dimensional vision of the original teddy bear.

The brilliance of a hologram is in allowing you to create a code for the three-dimensional image on a two-dimensional plate. Loosely speaking, you can think of the density of light and dark bands as representing depth in the missing dimension. In other words, a dense band of darkness encodes a perpendicular distance close to the plate, while a lighter band represents something much further away. 't Hooft and Susskind's holograms store the missing dimension in a very similar way. You feel three dimensions as opposed to two because that is how your brain has chosen to decode the light and dark bands. It has chosen to represent them as a third dimension of space and a little bit of gravity.

't Hooft and Susskind's conjecture is usually referred to as *the Holographic Principle*. To do it justice, we really ought to talk about it in the language of relativity and quantum mechanics. So, in other words, we really ought to be talking about *quantum* gravity in a *four*-dimensional *spacetime* and a *quantum* hologram on its *three*-dimensional boundary (with two dimensions of space and one of time). We can also imagine applying these holographic rules in a variety of other universes, including those that look completely different to our own. These are purely hypothetical worlds squashed and warped in wonderful ways, and they might even involve extra dimensions of space, beyond the three that we normally think about. But whatever the spacetime, we can try and play the holographic game. What you expect to get is two equivalent descriptions of the same physics: a higher-dimensional world with gravity versus a lower-dimensional boundary world without gravity. For example, in a world with six dimensions of space and one of time, you might talk about gravity in seven-dimensional spacetime and a hologram living on its six-dimensional boundary. The point is that the holographic principle is something you can think about whenever you think about gravity.

There can be no doubt about it: 't Hooft and Susskind's work should be hailed as a revolution in our understanding of quantum gravity. It has allowed us to recast old problems in a new and improved holographic language. An example of this is the information paradox. You may remember it from the Chapter 'Graham's Number'. Hawking was convinced that black holes lost information and that meant they violated the fundamental quantum laws. But taking into account holography, you realize that this can't be true. This is because there must be a way to encode the formation and evaporation of a black hole on the boundary of space. As this alternative lower-dimensional description does not include any gravity, we think of it as a much simpler quantum theory. It corresponds to a quantum dance of charged particles, pushed and pulled by ordinary simple forces, similar to those found in molecular interactions or nuclear physics. For holography to make sense, this boundary quantum theory should be mathematically consistent and physically well-behaved. Nothing should fail or break down when you describe events in this alternative language and therefore no information should be lost. Of course, this argument works only if holography is real.

So, is holography real?

This is the million-dollar question. We have no experimental proof that our world is a hologram. And even if there is a hologram, we aren't sure what it will look like. Of course, as 't Hooft and Susskind realized, black holes offer tantalizing hints that the hologram is there. But the reality of holography in our particular world remains a conjecture. There are, however, other worlds where holography is almost as good as proven.

These other worlds have been revealed by Juan Maldacena. Maldacena is a modern giant in the land of physics, a multi-award-winning professor at the elite Institute of Advanced Study in Princeton. In the last three decades, his contributions to our theoretical understanding of gravity and the universe have been second to none. Pound for pound, I would say that Maldacena is the greatest physicist alive today. Susskind calls him 'The Master'.

In the mid-1990s, Maldacena was still something of a newcomer. His reputation had been growing as a young PhD student from Buenos Aires working on string theory and the physics of black holes at Princeton University. A year after leaving Princeton, Maldacena was inspired by a talk he heard from the Russian physicist Sasha Polyakov at an international meeting in Amsterdam. Polyakov suggested that some aspects of nuclear physics in four dimensions could be connected to the strings of string theory, moving in a five-dimensional spacetime. The young Argentine made a series of brilliant connections and within a few months he unleashed an academic bombshell:

The Large N limit of superconformal field theories and supergravity.

It wasn't the catchiest of titles, but the content of the paper sent shockwaves across the academic community. It was the moment that Maldacena's rising star went supernova. He had discovered a holographic world, in which one dimension of space was nothing more than an illusion. It didn't matter that this world was very different to our own. What mattered was that it was mathematically simple enough for Maldacena to demonstrate exactly how the illusion worked. Holography now had to be taken seriously. The discovery of this strange and purely hypothetical world was a step change in our understanding of space and time at the most fundamental level.

Maldacena's world was like nothing you could really imagine, an exotic universe of strings and quantum gravity operating in ten dimensions of space and time, five of which were warped in a very particular way, five wrapped up like a sphere. On the boundary of that spacetime, Maldacena presented another theory – a theory without gravity capable of describing everything that happened in the interior. It was so important because he showed how and why the two descriptions should really be equivalent. He also figured out how to speak both languages fluently – the interior language and the boundary language – and, with the help of the formidable American physicist Ed Witten, began to write a dictionary between the two. Of course, 't Hooft and Susskind had speculated that holographic descriptions like this should exist, but that was about as far as they got. They were never able to hold up an example and say, 'Here is a universe with gravity, here is its hologram and here is the dictionary you need to go between the two.' But that is exactly what Maldacena *was* able to do. Although he was aware of 't Hooft and Susskind's earlier ideas, they were not at the forefront of his mind. It was Ed Witten who would connect Maldacena's theory to holography.

Ed Witten is another genius, a Fields medalist and one of *Time* magazine's hundred most influential people in the world in 2004. His father, Louis, was a theoretical physicist who would discuss his research with his precocious young son. He would always speak to him as if he were an adult and, despite his remarkable ability to understand his father's work, Ed would eventually be drawn away from physics. He would major in history at Brandeis University in Massachusetts and pursue a career as a journalist, writing articles for *The Nation* and *The New Republic*. But the lure of physics always remained. After enrolling on a masters programme at Princeton University and completing his doctorate, Witten would go on to become one of the founding fathers of string theory. 'He never does calculations except in his mind,' revealed his wife, Chiara Nappi, herself a physicist. 'I will fill pages with calculations before I understand what I'm doing. But Edward will sit down only to calculate a minus sign, or a factor of two.'

Maldacena's example of holography is known as the *AdS/CFT correspondence*. It describes a *duality* – two equivalent descriptions of exactly the same physical phenomena. On one side of the duality, we have AdS – shorthand for anti de Sitter, a warped higher-dimensional world where gravity is present as a fundamental force. On the other side we have a CFT. This stands for 'Conformal Field Theory', betraying a special mathematical property of the lower-dimensional hologram. There is no gravity on this side of the duality and yet, remarkably, it is capable of describing exactly the same physics. It does this through a weightless waltz of charged particles that are very similar to gluons, the fundamental force carriers that glue together atomic nuclei. The set-up was genuinely holographic: you could even think of the gluons as living on the boundary of that spacetime, on the outer wall of anti de Sitter space.

In his original paper, Maldacena gave a very compelling argument in support of this AdS/CFT correspondence, but he wasn't able to offer a strict mathematical proof. As time passed, however, his claim was tested again and again and again. Some physical quantities could be calculated accurately on both sides of the correspondence – on the one hand, using gravity and spacetime, and on the other, using the hologram. The results have always matched, and now there is no longer any room for reasonable doubt: the AdS/CFT correspondence really is a concrete example of the holographic principle in action. We can now imagine worlds – albeit warped anti de Sitter worlds – in which gravity and a dimension of space can be eliminated by a holographic trick.

But what about us? Do we really exist inside a hologram? This is a far more difficult question. We do not live in five-dimensional anti de Sitter space, so we cannot appeal to Maldacena's magic. But in our universe, black holes do funny things. Their entropy grows with the area of their boundary rather than their volume. Information looks as if it should be stored on the event horizon as opposed to the interior of the black hole. It is as if our world is telling us it is holographic, but keeping the hologram a secret, at least for the time being. If the hologram is there, you are experiencing gravity and certain dimensions of space because of the quirky way in which you decode that hologram. Look around you. Look left and right, forward and back, up and down. If our holographic expectation is right, one of those dimensions can be packaged into something completely different. The moment we free ourselves from our gravitational inhibitions, we no longer need to talk about three dimensions of space. Two will do us just fine.

It reminds me of the allegory of Plato's cave, and the prisoners who had been forever trapped inside. Chained up with a fire burning behind them, they would stare at the shadows that formed on the wall. To them, this was everything, their entertainment, a flattened perception of all there was. But Plato argued that with philosophy and ideas, the prisoners could break free from their chains. They could see beyond the shadows on to the puppets that were casting them. But I now think that Plato was underestimating the shadows. In a holographic world, the shadows are as real as the puppets.

The holographic principle is the most important idea to have emerged in physics in the last thirty years. It has led to breakthroughs in our understanding of the gravitational force, not least a resolution of the black hole information paradox and deep insights into the nature of quantum gravity. It has helped us to better understand the microscopic embrace of quarks and gluons as they pull together in a subatomic world. But more than all of that, it has challenged our perception of reality, our concept of the space that surrounds us. It has led us to ask if it really exists, or if it is just an illusion.

For us, this illusion is the legacy of leviathans, the largest and most magnificent of our fantastic numbers. With googolplician doppelgängers and black hole head death, with TREE(3) and the unreachable conclusion of the Game of Trees, we have told the story of entropy and quantum mechanics, of gravity and the mysterious physics of black holes, the cosmic oubliettes. These are the same ideas and concepts that underpin the holographic truth. They betray the terror of dimensional redundancy, of an alternative reality trapped on the boundary of space. We are the shadows on the wall.

Now, as we finally transition from large to small, from the big numbers to the little numbers, you should expect the unexpected. The little numbers will hint at symmetry and beauty but, in the end, they will leave us in despair. You should brace yourself for tales of a universe that shouldn't exist, that should have been crushed into oblivion the moment it was born. Our universe. Our unexpected universe. I cannot speak for you, my friend, but this worries me more than the shadows. It worries me to think that everything I know should never have existed: me, my family, my closest friends. This book should never have existed and yet, somehow, you're reading it, right now, in a moment that might never have arrived.

Little Numbers

Zero

A BEAUTIFUL NUMBER

I was finally starting to get excited. Liverpool Football Club had won twenty-six of their first twenty-seven games of the Premier League season. Jürgen Klopp, their charismatic German coach, had labelled his squad the 'mentality monsters', such was their ability to win game after game, even when the odds were stacked against them. This was no more evident than away at Aston Villa, in the face of a passionate home crowd, on a dreary November afternoon. Klopp's team were one–nil down with three minutes to go, but they came out victorious, as Sadio Mané, the Senegalese international, scored the winner with the very last touch of the ball. As the wins kept coming, the pundits were convinced: in 2020, Liverpool Football Club were going to win the Premier League.

I have supported Liverpool ever since I was a young lad growing up on the outskirts of the city. As a teenager standing on the Kop, one of the most famous terraces in world football, I had twice seen them lift the title. But both times had been over thirty years earlier. The decades in between had been barren in comparison, filled with crushing disappointment as my team fell consistently short in the league, more often than not to our greatest rivals from the neighbouring city of Manchester. So, despite the commanding lead that Liverpool held at the top of the table deep into the Premier League season, I could not bring myself to take anything for granted. I needed someone to crunch the numbers.

My friend Dan is an astronomer. He also supports Liverpool and will sometimes use my season ticket if I can't make it to a game. However, unlike me, Dan has a number of transferable skills and has put together a clever model for predicting the outcome of football matches. I asked Dan to perform a *million* simulations of the few remaining months of the season, just to set my mind at rest. When the results came back I was relieved. Dan's model had predicted Liverpool to win the Premier League on 999,980 occasions, Manchester City on 19 occasions, with Leicester City winning it just once.

What Dan had created was a kind of multiverse, a million parallel worlds containing a million Premier League tables. In almost every part of that multiverse, Liverpool would ultimately be crowned champions, so I was confident the three-decade drought would soon be over. However, I could not be *certain*. There were still some corners of the multiverse where Liverpool would capitulate and the title would find its way to Manchester or Leicester. Of course, this unhappy outcome (for me) would be most unexpected. Dan's multiverse predicted it would occur with a probability of just 0.00002 or, in other words, it was a 50,000–1 shot.

In the end, Liverpool won the Premier League, but not without an enormous scare. In March 2020, just two games away from victory, the season was suspended as the devastating coronavirus outbreak spread across the UK. During the strict national lockdown that spring, no one was sure when things would get back to normal. Football took a back seat and, as a Liverpool fan, I began to wonder if we were living in an unexpected and improbable corner of Dan's multiverse.

One thing is for sure, when you step away from football and into physics, you will find yourself in a very unexpected place. In the multiverse of physical worlds, our universe is in the most improbable of corners. The surprises begin with the discovery of the Higgs boson at CERN but run deep into the bubbling broth of the vacuum of space. The truth is that our universe is plagued by inexplicably small numbers and terribly improbable outcomes. These demand an explanation. If Liverpool had failed to win the league when the probability of failure was just 0.00002, you'd be curious to know what went wrong. A deadly virus, perhaps? So, it is in physics, when faced with tiny numbers and the unexpected properties of our universe, that we start asking questions. What is it that makes the Higgs particle so ridiculously light? Why is the bubbling broth of the vacuum so inexplicably gentle? These are the tales of the unexpected, a quest to understand the smallest numbers in physics, to make sense of an improbable universe that should never have existed.

We'll begin with zero. In absolute value, it doesn't get any smaller than that.

Zero is symmetry.

We can glimpse at the connection between symmetry and zero by imagining the accounts of a large organization. The books show millions of dollars going in and out on a regular basis. A careless look at a few individual transactions suggests that the incomings and outgoings are more or less random, save for their overall scale. But there is also something strange about these accounts: at the end of each quarter, the chief accountant reports a profit of exactly zero. In other words, the organization always breaks even. Generically, this wouldn't happen. Typically, we'd expect the books to reveal a profit or loss measured in millions. It's like trying to balance a herd of African elephants against a herd of Indian elephants – the scales are bound to tip one way or the other. A *zero* in the profit and loss of the company accounts would indicate a perfect *symmetry* between incomings and outgoings, and that merits an explanation. In this case, it might be that the organization is a charitable trust, committed to good accounting and determined to feed all its profit back into good causes. The point is that a vanishing value, be it in accountancy, physics or a herd of elephants, doesn't happen by accident. There is always a good reason and it normally involves a symmetry.

Symmetry is nature's ideology. The interactions of subatomic particles, the building blocks of everything you see, are governed by the symmetries of the Standard Model of Particle Physics. The twentieth century taught us that the clues to understanding physics are often in the smallest of nature's numbers. Whenever we see a zero or something unexpectedly small, we start thinking about the symmetries that could be responsible.

So what is symmetry?

Symmetries are a turn-on. And I don't just mean for an overexcited physicist. As humans, we find symmetry physically attractive. Research has shown that a good balance between the left and right side of your face is often considered to be beautiful. This is usually explained by the Evolutionary Advantage Theory. Our genes are designed to develop a symmetrical face, but other things can get in the way of that: age, disease, parasitic infection. These are all indicators of poor health. We are therefore attracted to facial symmetry because, from an evolutionary perspective, we want to mate with someone healthy.

Symmetry has also inspired artists through the ages. We see this in the bilateral and rotational symmetries of tribal images as well as in the imperial patterns that adorn the Alhambra, the spectacular fourteenth-century Islamic palace in the Spanish city of Granada. When the Islamic artists were conjuring up their decorations on the floors and walls of the Alhambra, they created different shapes and patterns betraying different kinds of symmetry. These could be grouped together in the way they combined the familiar symmetries of reflection and rotation as well as in the less familiar symmetries of translation and glide reflection.[1]

To see how the patterns of the Alhambra can be classified by symmetry, consider the tiling shown in the figure below, taken from the Patio de los Arrayanes (Court of the Myrtles):

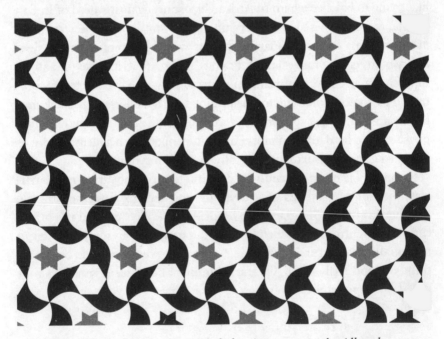

The dancing bats in the Patio de los Arrayanes in the Alhambra.

To me, the pattern conjures up an image of bats dancing in threes against a starlit sky. But the real beauty is in its symmetry. You can see how it possesses translational symmetry by jumping from left to right or along the diagonal. The image is also filled with threefold rotations. For example, if you rotate through a third of a circle (120 degrees) about the centre of any star, the image doesn't change. You can do the same about the centre of one of the white hexagons or even about the point where three batwings come together and touch. In other words, you have a triplet of threefold rotations alongside a pair of translations. This particular combination of symmetries is known to mathematicians as the p3 group. In homage to the bats of the Alhambra, we shall call it the three-way waltz.

Other patterns may also contain the same symmetries as the three-way waltz, or they may contain different ones, making them mathematically more distinct. By grouping together rotations, reflections, translations and glides in a variety of different ways, we can easily imagine a never-ending Alhambra of mathematically distinct patterns, an infinite number of patios each adorned with its own bespoke symmetry group. The different groups that characterize these patterns are known collectively as the wallpaper groups, for obvious reasons. But here is something unexpected: the early Islamic artists only ever reproduced seventeen of these. That doesn't seem like very many. In fact, if we were to look across all cultures, it would seem that no one has ever produced anything beyond these seventeen patterns. At first glance this is strange. After all, you could imagine combining rotations, reflections, translations and glides in a vast number of different ways, so you would expect there to be a very large, perhaps even an infinite number of wallpaper groups. Why would the great artists of history only ever touch upon seventeen of them? Well, it was certainly not through lack of imagination. It turns out that there is a limit to our mathematical beauty. Because of the need to repeat the pattern, you can only ever find seventeen different combinations that sew together in just the right way. You can prove this with the so-called *magic theorem of mathematics*.[2] It would seem that the artists of Islam were creative enough to capture everything there could ever be.

What this teaches us is that symmetries are really special. They don't just allow any old pattern, and this is true whether we are talking about the art of the Alhambra or the cosmic art of our unexpected universe. If something is special or unexpected, chances are that symmetries are responsible. Since they are the key to unlocking the mysteries of the universe, we probably ought to decide what a symmetry really is. When I asked my eldest daughter what she thought of when I said the word 'symmetry', she said a square. I thought this was a pretty good answer. After all, a square has some very well-defined mathematical beauty. If you rotate it through ninety degrees about its centre, it will look exactly the same. It will also stay the same if you flip it along a diagonal or across a line passing through the centre of opposite sides. Ultimately, that's what we really mean by a symmetry: you act on something in a non-trivial way and that action leaves it unchanged. For example, for a human face, the action is to perform a reflection and, if it is truly beautiful, the face will remain unchanged. For the tiles of dancing bats in the Alhambra, the relevant actions are the translations and threefold rotations.

And what about zero? Is there an action that leaves it unchanged? If you want to think of zero as a real number, then one possible action you could perform is a sign flip. In other words, you send five to minus five, minus TREE(3) to TREE(3), and so on. A sign flip will generically take you to a different place on the number line, with one exception: zero. When you flip the sign of zero you stay at zero. In other words, zero is the only real number that is *symmetric* under a change of sign. You could extend the idea to complex numbers. Now you can say that zero is the only complex number that stays put when you rotate the argument. Of course, the connection between zero and symmetry runs much deeper than a few mathematical tricks. As we will see, an unexpected zero is nature's way of telling you that there is an underlying symmetry in the fabric of our physical world. And because there is beauty in symmetry, that should mean that there is beauty in zero.

And there is.

But our ancestors didn't always see it this way. There is another side to zero that I also need to tell you about, an historic tale of suspicion and mistrust. The problem was that, in zero, ancient scholars saw deep into the void. They saw nothing, the absence of God and the essence of evil. As the philosopher Boethius wrote while awaiting execution in 524:

'Can, then, God do evil?'

'No.'

'Then evil is nothing, since it is beyond His power, and nothing is beyond His Power.'

To his medieval mind, zero was not the object of beauty that I see. It was the Devil himself.

A HISTORY OF NOTHING

Now is the time for the devils.

Now is the time to tell the story of a beautiful number from the very beginning, to reveal the truth of its difficult journey through the paranoia of human history. We shall follow it step by step, moving from one ancient civilization to the next, from Mesopotamia to Greece, from India to Arabia, until we finally arrive at the devils and accountants of Western Europe. Each of these will tell their story: in some cases, a celebration of zero but, more often than not, a withering contempt. The history of nothing begins in the Fertile Crescent of modern-day Iraq. It begins with the birth of numbers.

It was here in Sumer, in ancient Mesopotamia, that the world's oldest civilization was born more than six thousand years ago. The ancient Sumerian city states of Uruk Lagash, Ur and Eridu lay nestled in between the rivers Tigris and Euphrates, in areas of lush farmland. Here, as in Egypt, civilization seemed to demand mathematics before it demanded prose, with the earliest records corresponding to numbered inventories as opposed to the written word. Or to put it another way: the accountants came first. In the story of zero, they would also come last.

From around 3000 BC, the Sumerian accountants would mark their inventories on clay tablets. If they wanted to record five loaves and five fish, they would mark five pictures of a loaf and five pictures of a fish. Their first great intellectual leap came from separating the number from the object they were counting. In other words, they would represent five loaves with a *numeral* for the *number* five alongside a symbol for the bread. If they wanted to describe five of something else, they realized they could keep the same numeral and trade the object symbol for a fish, or a jar of oil, or whatever else they were interested in. The Sumerians had developed the idea of an emancipated number, existing in its own right and independent of whatever it is being used to count. It is easy to take the emancipated number for granted as it is so ingrained into modern thought, but to the earliest civilizations it was intellectually new and supremely powerful.

With this breakthrough, the Sumerians began to develop a number system loosely centred around the number sixty with bespoke symbols for 1, 10, 60, 600, 3,600 and 36,000. We do not really know why they chose a system that was predominantly sexagesimal. The most popular theory, dating all the way back to Theon of Alexandria (AD 335–405), is that sixty was chosen because it had so many divisors. Whatever the reason, a legacy of this sexagesimal thinking still survives today, as we count sixty seconds in every minute and sixty minutes in every hour.

There was no subtlety in this early system of numbers. The Sumerians would simply pile the symbols up until they got to the number they wanted. For example, if they wanted to express the number 1,278, they would pile up two copies of 600, a 60, a 10, and eight copies of 1. It wasn't especially efficient. That all changed around 2000 BC when the Mesopotamian mathematicians made their next great intellectual leap: they began to appreciate the importance of position. The Sumerians and their Babylonian successors began to develop a new number system based on just two numerals: the wedge T generically meaning one and the hook < generically meaning ten. Crucially, however, the relative position of these made a difference to the overall meaning. For example, consider the number 56. This would be written with five copies of a hook (ten) and six copies of a wedge (one):

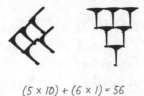

$$(5 \times 10) + (6 \times 1) = 56$$

That doesn't seem especially clever. But suppose we move two of the wedges to the far left of the page:

$$(2 \times 60) + (5 \times 10) + (4 \times 1) = 174$$

Rather than representing a pair of ones, the Babylonian mathematicians would now interpret these as a pair of sixties and the number would become 174. What they had developed was a sexagesimal system where the relative location of the numerals determined how many powers of sixty you should count. Here is another example:

$$(1 \times 60^2) + (3 \times 60) + (4 \times 10) + (2 \times 1) = 3822$$

This was the cleverest number system the world had ever seen. It was supremely efficient, the positional code significantly reducing the number of symbols needed to express a number. But something was missing. Or perhaps more precisely, *nothing* was missing. I'll explain with a story.

An ancient Babylonian mathematician is summoned by a priest and asked to write down the number of offerings that were submitted to the temple. A bag of grain. A wooden carving. Ivory, silk and precious metals. He counts them all and finds that there are sixty-two offerings in total. That is $62 = (1 \times 60) + (2 \times 1)$, so he carves out the following symbols on to a clay tablet and hands it to the priest:

In the week that follows there are many more offerings, far more than had gone before. More jewels, more gold, more wine and more food. Once again, the mathematician is asked to count them and make a record on another clay tablet.

Once he has finished counting, he gathers his stylus and carves out the following:

The priest is furious. Clearly the mathematician is a crook. This week's offerings were far more abundant and yet here he is recording exactly the same number. He is not to be fooled and orders the mathematician to be executed. As he is dragged towards his death, the mathematician protests his innocence. He had counted 3,602 offerings that week and it was indeed far more than the 62 of the previous week. But in a sexagesimal system $3,602 = (1 \times 60^2) + (2 \times 1)$, so it could only be written as he had proposed. The priest, like much of Babylonian society, was not well versed in the subtleties of this new positional system of numbers. As far as he was concerned, the mathematician had written down the same thing twice. He had tried to trick him. In the end, nothing could have saved the mathematician. By nothing, I mean zero. A zero could have saved him.

In a sexagesimal system, we can write $3,602 = (1 \times 60^2) + (0 \times 60) + (2 \times 1)$, and so it should really be labelled with a one $\intercal$, a *zero,* and finally a two $\intercal\intercal$. This is what distinguishes it from $62 = (1 \times 60) + (2 \times 1)$, labelled with a one $\intercal$ and a two $\intercal\intercal$. But the old Babylonians would only mark a zero by leaving an empty space, and they didn't always make it especially large. By their reckoning, any ambiguity had to be understood from the context. As we saw with the lamentable tale of the mathematician in the temple, such a system can easily fail. On each of the tablets, the priest could not tell if the first symbol was followed by a meaningless space or a meaningful zero.

The positional number system of Old Babylonia was a brilliant piece of mathematics but, in the absence of a symbol for zero, it was fundamentally flawed. By around 1600 BC its use had gone into decline, and the system would remain dormant for more than a millennium. Its resurrection came after Alexander the Great and his armies of Macedon had conquered Mesopotamia in the third century BC. Alexander died suddenly at the height of his power in the Palace of Nebuchadnezzar in Babylon, aged just thirty-two years old. In the bloody years that followed, his empire was divided and a huge portion covering much of Asia fell into the hands of his general, Seleucus. It was during this

Seleucid period, from 321 BC until the Roman conquest in 63 BC, that
the Mesopotamian mathematicians made their third great intellectual
leap. They rediscovered the brilliance of their positional numbers and
flavoured them with a remarkable new ingredient:

Whenever you saw this symbol in a number, it counted an empty
column of 60s or 3,600s, depending on where it was placed. It was a
zero; not an emancipated zero but a placeholder. If our ancient mathematician had known it, he could have been spared the wrath of the
priest. He could have written 3,602 far more distinctly:

(1 × 3600) + (0 × 60) + (2 × 1) = 3602

The new symbol for zero removed some of the ambiguity that had
plagued the positional number system. It offered the mathematicians
and astronomers of ancient Babylon unprecedented calculational
power, even if it didn't catch on among the wider population. Strangely
enough, the scholars only ever placed the symbolic zero at the beginning or the middle of a number, never at the end, so some ambiguity
still remained. Nor would it ever be found on its own – it would never
be emancipated. The symbol had originally been used to separate sentences, rather than numbers, suggesting it might really have indicated
a space, as opposed to a number in its own right. Even so, the Babylonians had staked their claim for the invention of zero, at least as a
rudimentary placeholder.

Rival bids for the first zero come from the Maya, in Mesoamerica,
and, of course, the ancient Egyptians. The Mayan zero took the form
of a shell, or sometimes the head of a god, his hand pressed thought-
fully across his chin. Although it probably came earlier than the
Babylonian zero, the Mayan version was neither an emancipated
number nor indeed a placeholder. Instead, it was tied to the keeping

of time, helping to measure the number of days, months and years since the Mayan *zero* day, the mythical moment of creation dated 11 August 3114 BC in today's calendar. The Egyptians never used zero in any of their numbers, but they did use *nfr*, written as ⚭, to represent a vanishing balance in accounts or the ground level at pyramid construction sites. In their ancient language it meant 'good', 'complete' or even 'beautiful'. This resonates strongly with our theme of zero as an avatar of symmetry and beauty.

Neither the Mayan nor the Egyptian zero made it far beyond the shores of their civilizations. However, the Babylonian zero did, taken to Greece in the years that followed Alexander's conquest, along with gold, and the women and children taken as slaves. The Greeks wrote their numbers with an alphabetic code. They had letters corresponding to certain numbers such as 1, 2 or 100, and took combinations of these to make other numbers such as 101 or 102.* But they never made any clever use of positions. Even so, when the Greek mathematicians discovered the Babylonian system, an elite few were smart enough to recognize its advantages, although they preferred to keep their admiration private. They began to perform their more complicated calculations with the imported system, before translating their results back into the old Greek style. As for the Babylonian zero, the Greeks certainly knew about it and eventually came up with their own symbol, ☉, which is tantalizingly similar to the one we use today. This is probably just a coincidence, as the symbol didn't make its way into any of the older Western numerals. The Greeks enhanced the Babylonian system by placing zeroes at the end of their numbers, but they never set zero free – they never recognized it as a number in its own right. Given the reputation of the Greek mathematicians, it is natural to ask why. At some level, it just didn't interest them. Greek mathematics was dominated by geometry, by tangible lengths and shapes, so it is hard to see where they would have found a role for zero. But it also ran deeper than that. The Greeks had a contempt for zero, a mistrust, that the West was only too willing to follow.

It was a question of philosophy.

---

* 1, 2 and 100 were respectively written as $\bar{\alpha}$, $\bar{\beta}$ and $\bar{\rho}$, and so 101 would be written $\overline{\rho\alpha}$ and 102 as $\overline{\rho\beta}$. A bar was written over the top of the letter to distinguish numbers from words.

The problems started with Zeno of Elea.[3] Zeno was a prominent member of the city's philosophical school, led by his tutor, Parmenides, who rejected the notion of change, arguing that the movements we see are just an illusion. He applied this to everything – a chariot in a race, an arrow through the air, the torrent of a waterfall. None of this motion was real. Of course, it seems absurd. We can see with our own eyes the diverse and changing landscape that surrounds us. But Zeno cooked up a series of paradoxes that seemed to prove our senses should not be trusted to reveal the truth. Although it wasn't immediately apparent, there was one particular paradox whose understanding and misunderstanding was intimately connected to zero.

We'll tell our own version of the story. Achilles, the greatest of all the warriors in Greek mythology, finds himself in a foot race with a tortoise. He is confident of success – after all, he has a top speed of 10 metres per second and no one has ever seen his languid opponent go faster than a tenth of that speed. He decides to give the reptile a head start and begins the race 10 metres behind. He is instantly up to his top speed, and within a single second he has reached the point where the tortoise started the race. But the tortoise is no longer there. It hasn't moved far, of course, just a single metre forward, but the reality is that Achilles still hasn't caught up. In a tenth of a second, Achilles has made up the missing metre, but the tortoise has moved on again, this time advancing another 10 centimetres. By the time Achilles has run those 10 centimetres, the tortoise has advanced another centimetre, and so on. With each new step, Achilles gets closer, but to catch up with the tortoise he would need an infinite number of steps. In other words, he would never catch up.

Zeno had baffled his contemporaries. Clearly Achilles would rise to the challenge and overtake the tortoise in a matter of seconds, but how could they refute Zeno's argument? They saw that the issue lay somewhere in the infinity of steps – and they were right. But to conquer the problem of infinity they would also need the mathematics of zero, which they didn't have. Zeno didn't care. To him, their failure proved our senses could not be trusted. It was a triumph for Parmenides.

Zeno died a violent death. He lived in the ancient Greek city of Elea, under the rule of a cruel and tyrannical leader called Nearchus. Zeno had conspired to overthrow the tyrant, but when the conspiracy was discovered he was arrested and handed over to Nearchus and his men. They demanded to know the names of the other conspirators but, even as they tortured him, he wouldn't give them up. He whispered that he did have a secret but Nearchus would have to come close if he wanted to hear it. As he leaned in, Zeno sunk his teeth into the tyrant and would not let go. He was stabbed to death. Some say that he took Nearchus's ear; others say it was his nose.

A century later, Aristotle – the father of Western philosophy – began to think about Zeno's paradox. He tackled it with legislation, declaring that, in nature, there could never be an infinite number. Zeno had tried to divide the race into an infinite number of pieces. By Aristotle's law, those pieces could not be real in and of themselves – they were just a figment of Zeno's imagination. In the end, the only reality was the continuum of the race, with Achilles running past the tortoise in one continuous movement.

Aristotle did acknowledge the *potential* for infinite numbers but argued that this potential could never be realized. To understand what he was getting at, suppose you are slicing up a chocolate cake. You can slice it again and again and, in principle, you could imagine slicing it for ever, an infinite number of times. However, in the real world, we know you will never reach this point. We can recognize the *potential* to reach infinity, but we also know you will never hold a cake with an infinite number of infinitely small slices. To put it another way, you can hold infinity in your mind but never in your hand. According to Aristotle, this was where Zeno had come unstuck.

With our modern understanding of zero, we can bridge the gap between Zeno's imagination and Aristotle's continuum. The point is that an infinite number of steps does not automatically mean an infinite time. You can sometimes get a finite time, as long as the steps are getting shorter and shorter, approaching *zero* as the number of steps approaches infinity. If we take a closer look at Zeno's paradox, we see that Achilles completes the first stage after 1 second, the second after 1.1 seconds, the third after 1.11 seconds, the fourth after 1.111 seconds, and so on, the increments getting smaller and smaller. By extrapolating the result through an infinite number of steps, we see that the total number of seconds is 1.1 recurring. This is mathematically equivalent to 1 + 1/9 of a second.[4] The paradox is resolved: not only does Achilles overtake the tortoise, he does so in less than 2 seconds.

Without a proper understanding of zero, this resolution was always going to be beyond Aristotle and the other Greek philosophers. Indeed, it would be more than two thousand years before Zeno's paradox was fully understood. For this, Aristotle must take some responsibility. His rejection of infinity was the first in a trinity of ideas that led to a deep-seated mistrust of zero in Western thought. In rejecting the infinite, Aristotle had also rejected the infinitely small, the vanishingly short steps in the limit of Achilles' race. But in the second part of his ideological trinity, he went further. He rejected the void, the emptiness of space and the essence of nothing. To the medieval minds studying his works, this was a rejection of zero.

It came about because Aristotle was at war with the atomists, rival philosophers who believed that matter could not be broken up indefinitely. Instead, they declared it to be made up of tiny indivisible pieces, or 'atoms', frolicking around in an infinite void. This gave them their own alternative take on Zeno's paradox: if matter couldn't be broken down again and again, how could Zeno break the race up into ever-decreasing steps? The atomist perspective was completely at odds with Aristotle's. He believed that matter was a single continuous fluid, contracting and expanding, changing between the four basic elements of earth, water, air and fire. In his model, the universe was divided up into concentric spheres: at the centre, the terrestrial spheres, where humans lived, and on the edges, the celestial spheres, glittering with heavenly bodies such as the Moon, the Sun, the planets and stars. The terrestrial spheres were a changing and corruptible landscape, separated into four layers: earth at the centre, then layers of water, air and eventually fire. Matter could transform from one form into another. When it was cold and dry, it became earth; when it was cold and moist, water; when hot and moist, air; and when hot and dry, fire. As it changed its form, it would move through the layers until it found its natural place, earth falling to the centre and fire rising to the edge.

Aristotle's universe didn't need a void. But the atomist universe did – it needed something for the particles to move around in – so

Aristotle set about dismantling the idea. He began by thinking about how solid objects fall towards the ground. When they fall through a dense medium, like water, he noticed that they fall more slowly than they would through a rare medium. He also claimed that heavier objects fall faster than light ones, no doubt thinking about stones and feathers falling through the air. On the back of this, he decided that the speed of a falling object must be proportional to a simple ratio:

$$\frac{weight\ of\ object}{density\ of\ medium}$$

The void was immediately in trouble. Because it had vanishing density, all objects were expected to hurtle through it at infinite speed, and the space between atoms would fill up infinitely quickly. This couldn't be allowed to happen and so the void couldn't be allowed to exist. Of course, a stone doesn't fall faster than a feather because of its weight but because of air resistance. This was the weakness in Aristotle's logic, but it didn't matter – the damage was done. To Aristotle and his followers, the void did not exist. There could never be infinity and there could never be zero.

Why did these ideas endure for so long? What was it in Aristotle's work that appealed so much to the scholars of medieval Europe? It was the third piece of his ideological trinity: his proof of God's existence. This came from the celestial spheres, made from a fifth element known as the aether. Unlike the four terrestrial elements, the aether could not change form – it was incorruptible. The aetherial layers spread outwards from the terrestrial layers, each of them spinning by different amounts. There was a layer for the Moon, a layer for the Sun and a layer for each of the planets, the wandering stars. Surrounding all of this was a final layer of eternal darkness peppered with twinkling lights. These were the fixed stars, moving as one at the edge of the material world. But where did all this movement come from? What was conducting the heavenly orchestra? Aristotle argued that for something to move something else must cause it to move. For example, you could imagine each sphere being moved by its larger neighbour: the lunar sphere moved by Mercury, Mercury moved by Venus, and so on. But if we do this, what happens when we reach the final layer of stars? Who moves that? Aristotle claimed that this movement came from beyond the material world. It came from the *prime mover* or, in other words, it came from God.

As Christianity swept its way through the Western world, it is easy to see how it was drawn to this philosophy. Although Aristotle had proven the existence of a secular god, Christians like St Thomas Aquinas were happy to take it as proof of their own. They embraced Aristotle's universe and came to believe that support for the atomists was to deny their god's existence. They rejected the void and they rejected zero.

But the story of zero continued. Like the Sun, it would rise in the East. Perhaps we should really speak of the rise of *śūnya*. This is the Sanskrit word for zero, but it also means void. Unlike the Christians with their fears of heresy, the Buddhists would embrace the void – it was at the heart of their spirituality. *Śūnyatā* was the emptiness of emptiness. A Buddhist would seek their liberation into the void through the power of meditation. Similar ideas could be found in other Eastern religions, such as Hinduism and Jainism.

Some say that zero was taken to India from Babylon in the years that followed Alexander's conquest; others say it grew from within, from the seed of the *Śūnyatā*. We do not know. What we do know is that India provided the root of our own zero. It is where we began to see the symbols that would pass through generations, arriving at the circular glyph we see today. But more importantly than that, it was in India that zero would finally be set free.

At some point in the middle of the first millennium, the Indians switched to a system of numbers closely resembling our own. Like the Babylonians, they made clever use of positions, but it was a decimal system, as opposed to sexagesimal. It's hard to know when the switch came about exactly on account of fraud. Many of the earliest documents were of a legal nature, certifying the granting of land to certain individuals. Since they were later used as proof of ownership in historical land claims, the dates on them were often forged.

There were those who took advantage of this to support the idea that Indian numerals did not appear until the ninth century. If any dates suggested they were older than that, the documents were dismissed as forgeries. This fanatical view stemmed from the work of George R. Kaye, an influential British scholar and orientalist from the early part of the twentieth century. Kaye had a dangerous agenda. He despised India and was determined to establish European supremacy in the realm of mathematics. By discrediting the earlier Indian documents, he could argue that the modern number system was not an Indian invention but had been imported into India from Greece or Arabia. Sadly, Kaye found plenty of support for his ideas among other British scholars, many of whom allowed their anti-Eastern bias to cloud their academic judgement.

Kaye's perspective has been widely discredited. While we are right to be wary of some documents, it seems unlikely that all of them are inscribed with erroneous dates, and most scholars now agree that our modern number system had emerged in India by the fifth century. This includes zero. We can trace its ancestry back to some leaves of birch bark discovered in 1881 by a peasant in the village of Bakhshali, in modern-day Pakistan. The bark contains mathematical text – methods for computing square roots and negative numbers – and a set of numerals, some of which are *almost* recognizable today:

List of numerals appearing on the Bakhshali manuscript.

Zero is represented by a dot, a direct ancestor of the circle we see in our own numerals. The date of the Bakhshali manuscript is a source of great confusion. Fuelled by his own prejudice, Kaye argued that it couldn't pre-date the twelfth century, but it is clearly much older than that. Analysis of the text suggests that it could be a copy of a more ancient work, perhaps as old as the third century. The manuscript is held at the Bodleian Library in Oxford, where three different samples were taken for carbon dating in the hope of settling the debate. It did anything but, each sample being dated to a different period in history: AD 224–383, AD 680–779 and AD 885–993.[5]

Zero was eventually set free by the great Indian mathematician and astronomer Brahmagupta. In the year 628, he wrote the Brāhmasphuṭasiddhānta, the 'correctly established doctrine of Brahma'. He was playing with negative numbers and, at the threshold of those numbers, he saw *śūnya*. He began to think about the meaning of sums and differences, of multiplication and division. If $3 - 4$ was a number, then why not $3 - 3$? Brahmagupta saw that zero was a true number, not just a placeholder but an honest player in the game of mathematics. The rules were simple: take any number, if you add or subtract zero you get the same number; if you multiply by zero, you get zero; and if you divide by zero . . . well, maybe not *everything* was simple.

When Brahmagupta tried to divide by his new number he started to get things wrong. For example, he declared zero divided by zero to be zero, but this isn't necessarily true. To see why, imagine there are two identical twins. They each take a shrinking drug and suddenly begin to lose size. In an instant their heights are halved, then halved again, and again, continuing for ever until they have shrunk to zero. As they both shrink at exactly the same rate, the ratio of their heights is always one. It never changes, so if and when both men have shrunk down to zero in the infinite future, their ratio must still be one. That must mean zero divided by zero is one, right? Well, that's not necessarily true either. What if the drugs had been taken by a giant and a dwarf? The giant is ten times taller than the dwarf at the start, and because they also shrink at the same rate, their height ratio also stays the same. It stays at ten. Follow it through as before and you might conclude that zero divided by zero is ten. But didn't we just prove it was one? The truth is it could be anything. It could be zero, one, ten, TREE(3) or even infinity. A ratio of zeroes is ill-defined on its own. You can take the ratio of two really small numbers and study the limit as they get smaller and smaller. That makes perfect sense mathematically but, as we have just seen, the final answer will always depend on how you approach the limit. Zero divided by zero is meaningless until you explain where the zeroes came from. How fast did you take the numerator to zero compared to the denominator?

When it came to dividing one by zero, Brahmagupta gave up. No wonder. As another Indian magician, Bhāskarāchārya, wrote in the twelfth century, a division like this yields *khahara* – the infinite – as unchangeable as Vishnu, the infinite Almighty. Eight hundred years later a division by zero would bring down the might of the American military. On 21 September 1997 there was a zero lurking deep inside the computer systems of the USS *Yorktown*, a 10,000-tonne missile cruiser stationed off the coast of Cape Charles, Virginia. In a single division, it took out the entire network, the propulsion failed and the ship was paralysed. According to Tony DiGiorgio, an engineer for the Atlantic Fleet and a self-described whistleblower, the *Yorktown* had to be towed to Norfolk Naval Base, where it lay dormant for two days. Atlantic Fleet officials denied this version of events, but they did accept that a division by zero had left the ship dead in the water for almost three hours. Zero might just be a number, as Brahmagupta realized, but whatever you do, don't divide by it, especially if the enemy are about to engage.

With zero now freed, it was ready to reach out across the world. In the early part of the fifth century, just as Brahmagupta was finishing off his masterpiece, the prophet Muhammad was ordering his followers to prepare for a pilgrimage to Mecca. Islam was beginning to spread across the Middle East. In the centuries that followed, it would continue to spread, expanding into a vast and magnificent empire, from Spain in the West to China in the East. Its vibrancy came from the veins and arteries of trade, along which there wasn't just a flow of goods but a flow of ideas: religion, of course, but also mathematics.

At the centre of this intellectual world was the House of Wisdom in Baghdad. The leaders of the Islamic caliphates had come to understand the importance of knowledge. They would send scholars on expeditions to collect texts from far-away corners of the empire. This was especially true of Caliph Al-Ma'mun, the most scholarly of the Abbasid caliphs, who reigned in the early part of the ninth century. It was during his tenure that the House of Wisdom would blossom into the greatest centre of learning the world had ever seen. One of its scholars was a brilliant Persian mathematician by the name of Muhammad ibn Mūsô al-Khwārizmī. Al-Khwārizmī is celebrated for *Al-jabr*, a compendium of mathematical techniques for solving equations and from where we take the word 'algebra'. This was one of the most important treatises in the history of mathematics. The ancient Greek obsession with geometry was replaced with a subtle form of mathematical reduction. Questions became equations, answers the roots of those equations, and algebra the magic that tied it all together.

By the time of al-Khwārizmī's work, the Indians were no longer using dots for zero but circles. The Arabs had learned of zero and the rest of the Indian numerals around fifty years earlier, after a diplomatic visit from the Indian province of Sindh to the court of Caliph Al-Mansur in Baghdad in 773. The Sindh embassy had brought along a copy of Brahmagupta's book as a gift to the caliph. When al-Khwārizmī came to study it decades later, he immediately realized its importance. He began to unpick the rules of Indian arithmetic, including zero, and developed algorithms for long addition, subtraction, multiplication and division. In fact, the word 'algorithm' comes from *algorismus*, a Latin corruption of the name al-Khwārizmī. Despite the Indian origin of today's numerals, al-Khwārizmī's legacy is such that they are now commonly referred to as Arabic. He had taken an uncut Indian jewel, polished it and held it aloft so that it sparkled brightly throughout the Islamic world and eventually beyond.

Al-Khwārizmī used the word *ṣifr* to describe zero. It is the root of the word we use today. *Ṣifr* was a direct translation of *śūnya* – the void or emptiness in such conflict with the teachings of Aristotle. The

The number 270 instantly recognizable on an inscription in the Chaturbhuj Temple in Gwalior, some 250 miles south of Delhi, dated to the ninth century.

Muslims certainly knew about Aristotle and his proof of God's existence. Why didn't they reject *ṣifr*? Why didn't they condemn zero, as
they would in the West? The truth is there were those who were starting to doubt Aristotle. At the turn of the tenth century a new school
of Islamic theology was beginning to develop. It was founded by Al-
Ash'arī, a Sunni, who rejected Aristotle in favour of his deadly rivals,
the atomists. It suited his idea of occasionalism, a radical attempt to
impose God's omnipotence on all of nature. Occasionalism declares
that all events are caused by God, from a bouncing ball to a human
thought. Time is broken up into a series of accidents – each of them the
will of God – and matter is broken up into atoms, the victims of those
accidents. In every discrete moment, God wills new accidents to happen, and the atoms arrange themselves accordingly. In a way, it's a
philosophy that resonates with quantum mechanics. The motion of
atoms is not deterministic. To the Asharites, it is fixed by the will of
God; in quantum theory, it is fixed by measurement.

The most celebrated member of the Asharite school was Abu Hamid
al-Ghazali, considered by many to be a *Mujjaddid*, a figure who
appears once a century to renew the faith of the Islamic people. In his
teaching, al-Ghazali condemned Aristotelian thought and similar
ideas that contradicted God's omnipotence, declaring that anyone
who followed them should be put to death. His influence was such
that it sparked the beginning of the end for natural philosophy in
medieval Islam, in favour of a religious hard line. And yet, by embra-
cing the atomists and the interstitial void, he allowed *ṣifr* to prosper.
Zero, it seemed, was approved by Allah.

In just seven years at the beginning of the eighth century, the
Umayyad caliphate spread relentlessly through the Iberian Peninsula.
They founded Al-Andalus and opened up a channel for Islamic knowledge to pass into Western Europe. That said, it was never an easy
frontier. The Christian and the Islamic world were often at war – from
Charlemagne's raids into northern Spain in 778 to the Eastern Cru-
sades of the eleventh, twelfth and thirteenth centuries. For much of
this time, the Christians were still using Roman numerals and had
little interest in the heresy of zero. They were committed to Aristotle,
his rejection of the void and his proof of God's existence. Zero challenged that. It challenged their faith.

The tide began to turn towards the end of the twelfth century, when Guglielmo Bonaccio, a customs official from Pisa, was sent to the Mediterranean town of Bugia in Algeria. He decided to take along his son, Leonardo. The Arab world was an intellectual melting pot and, if nothing else, his son could learn to use the abacus there. But Leonardo learned so much more. He fell in love with Arabic mathematics and Indian numerals, and it was a love affair that would make him eternally famous. You probably know him by another name.

Fibonacci.

He became known as this by accident. Leonardo signed his work 'filius Bonacci', meaning 'son of Bonaccio', which scholars would later mistake for a surname, Fibonacci. But he was never known as Fibonacci in his lifetime. He was known as Bigollo, which probably meant traveller. It was a suitable nickname because Fibonacci roamed widely – through Sicily, Greece, Syria and Egypt – gathering knowledge from everywhere he went. At the turn of the thirteenth century, when he was around thirty years old, he decided to settle down, returning to Pisa to work on a masterpiece. Two years later, in 1202, it was published: *Liber Abaci*. It was a treatise on the mathematics he had learned in the Arab world: on algebra and arithmetic, on the mathematics of trade, and on the marvellous Indian numerals he held in such great esteem. At the beginning of his first chapter, he wrote:

These are the nine figures of the Indians

9, 8, 7 ,6, 5, 4, 3, 2, 1

With these nine figures and with this sign 0, which in Arabic is called ṣifr, any number can be written.

Notice the segregation. Fibonacci speaks of zero as a 'sign', separate from the other nine 'figures'. Of course, he would have known about the work of Brahmagupta – of the emancipation of zero – but he could not bring himself to place it as an equal alongside the rest of the Indian numerals. It was just too eccentric. As enlightened as he was, Fibonacci was obviously still nervous around this number. But in the end, it didn't matter. This was the moment that zero and the rest of the Indian numerals broke through the defensive line. They had entered Christendom.

Much of Fibonacci's book is devoted to the mathematics of trade, using the Eastern algorithms to calculate profit and interest or to convert currencies. Despite the obvious advantages, the European traders were slow to take them on. Many still preferred to work with Roman numerals, calculating with an abacus or a counting board of beads and pebbles. A contest began to emerge between the *abacists*, who clung to the old ways of banking, and the *algorists*, who embraced the calculating power of Eastern mathematics.

The ordinary man did not trust this mystical import from the East, and neither did the authorities. In the city of Florence in 1299, the use of Indian numerals was banned to prevent fraud. After all, you could easily change a zero into a six or a nine. But the ban didn't stop the algorists. They continued to use the Indian numerals in private, summoning the spirit of al-Khwārizmī as they calculated. At first, they were dismissed as unchristian, spending more time on their algorithms than on prayer. But, as ever, commercial pressures told, and the authorities relented. Zero, and the rest of the Indian numerals, were just too powerful to ignore. They were destined to prosper.

Even the Church seemed ready for a change. In the thirteenth century, the Bishops of Paris issued a series of condemnations, lists of heretical teachings that could result in a student being excommunicated. This included the words of Aristotle, the man who had inspired the likes of St Thomas Aquinas with his proof of God. The bishops began to see challenges to God's omnipotence in Aristotle's ideas, just as the Muslims had a few centuries earlier. In the condemnation of 1277, Bishop Étienne Tempier considered the question of moving the heavens. Aristotle had said they could never be moved in a straight line as this would leave behind a vacuum, filled with the void he had emphatically rejected. To Tempier, this was an obvious heresy. God could do anything He wanted. He could move the heavens however He pleased; He could create a vacuum. Who was Aristotle to declare otherwise?

Aristotle's hold over Christian philosophy remained strong in places, but his influence was starting to crumble. If Christians could accept the void, they could also accept zero. However, it wasn't the Bishops of Paris who effected lasting change and the acceptance of zero. It was the accountants.

They invented double entry.

In a way, it's an underwhelming end to the history of zero, but this is how it finally won. Double-entry bookkeeping was introduced to account for the growing complexities in trade. The oldest medieval record of its use comes from the treasury of the Republic of Genoa in 1340. The system was simple but ingenious. In one column you tallied up your credits, in another your debits, and if everything was accounted for, the difference would be zero. It was a system that played to the strengths of the algorists, with their positive and negative numbers balanced on either side of an emancipated zero. In 1494, the father of accountancy, a Franciscan friar named Luca Pacioli, summed up the method in his legendary textbook on practical mathematics. He pinned everything with a number – debit, credit, and even the vanishing balance. There was no longer room for a reasoned debate. It was clear that zero had triumphed, not with a bang or the violent overthrow of religious ideals, but through stealth and a merchant's need to balance the books.

ZERO IS SYMMETRY

What is zero? Our ancestors said it was the void – an emptiness – cursed on the West by the absence of God and blessed to the East by its silent perfection. Perhaps you would say that zero is just a number, like one, two or Graham's number. But then I would have to ask: what is a number? Before the ancient Sumerians emancipated their numbers, they were only ever considered alongside something else: five loaves of bread, five fish, five jars of oil. The breakthrough came when the Sumerians identified a common thread in each of these collections: the emancipated number five. The link between numbers and the things that they count is a hard one to break. Is the five that counts the loaves really the same five that counts the fish?

The question really took off in the late nineteenth century when mathematicians like the troubled German Georg Cantor began to think about collections of objects – they began to think about sets. As we'll see in the chapter 'Infinity', set theory grew from Cantor's religious quest to step into infinity, to reach high into the infinite heavens. But it was another German, Gottlob Frege, who first started using sets to think about ordinary numbers like zero, one, two, three, and so on – numbers we usually refer to as the *natural numbers*.

When we talk about the set of five loaves and the set of five fish, it's obvious that the two sets are connected in a very simple way: every loaf can be paired with a single fish, and every fish can be paired with a single loaf. This tidy arrangement is what mathematicians call a one-to-one map, or a *bijection*. We can also find one-to-one maps between five loaves and five jars of oil, or five American presidents, or five pop stars in a boy band. All these five-numbered sets are connected. If we want to use set theory to describe the number five, which of them should we use? Frege felt that none of them was special. When it came to picking out the number five, he argued that there was no good reason to pick five American presidents ahead of five loaves of bread or any other five-numbered set. In the interests of diplomacy, he declared the number five to be all these sets put together. In other words, it was the set of all five-numbered sets!

You can even find zero in this formalism. It is the set of all sets with nothing in them. What is a set with nothing in it? Well, there is only one such set – the empty set! It's a perfectly consistent idea. For example, we could define the empty set as the set of square numbers that are prime or the set of dogs that are cats.

Frege began to develop the foundation of arithmetic in this new set-theoretic language, but as the second volume of his work went to press, a bombshell arrived at his home. It came in the form of a letter from the British philosopher and polymath Bertrand Russell. As ever with Russell, the letter was one of uncompromising brilliance, destroying an entire body of work with a single detonation. Frege's idea assumed you could always talk about the set of all sets with some characteristic property. That was why he was comfortable using the set of all five-numbered sets to represent the number five, or the set of all ten-numbered sets to represent the number ten. But defining large sets in such a cavalier way is fraught with danger. As Russell enquired, 'What about the set of all sets that do not contain themselves?'

In order to show you what Russell was getting at, let me tell you about a barber I know called Giuseppe. Giuseppe makes a good living by shaving all the men who do not shave themselves. When I first found out about this I started to wonder: who shaves Giuseppe? Perhaps he shaves himself. No, I thought, that can't be right, because he only shaves men who do not shave themselves. OK, so that must mean he doesn't shave himself. Well, that can't be right either: if he doesn't shave himself, then he must be shaved by Giuseppe.

But he is Giuseppe!

Russell's question to Frege was loaded with a very similar dynamite. Despite the damage he had done to Frege's proposal, Russell would try to resurrect some of his ideas in ways that avoided the paradox. He still thought of numbers in much the same way, collecting together all the sets of a given size. He just couldn't identify those collections as sets in their own right. It turns out there is a much simpler and more economical way to think about the natural numbers using sets and it relies on one number in particular: zero.

What set should we identify with zero? We have already figured that out. The obvious choice is the empty set – the set with nothing in it. It is useful to think of this in terms of an empty box. If we want to generate other numbers, we need boxes that aren't empty. For the number one, we need to fill a box with a single object. What object should we use? Well, at this stage, the only things we know about are zero and empty boxes. So we might as well fill this new box with an empty box and call the whole thing 'one'. In set-theoretic language we say that one is the set that contains the empty set. What about two? To build a box for the number two, it should contain two distinct objects. As it happens, we now have two objects to play with – the boxes we

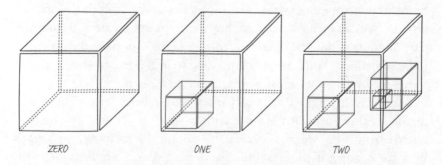

ZERO   ONE   TWO

*Constructing the natural numbers: ZERO is the empty set,
shown here as an empty box; ONE is now the box containing an
empty box; TWO is the box containing ZERO and ONE; and so on.*

identified with zero and one. All we have to do is put them both in the new box and call the whole thing 'two'. In other words, two is the set that contains the sets for zero and one.

We can carry on like this: three is the set that contains zero, one and two; four is the set that contains zero, one, two and three; and so on, sailing all the way past TREE(3) and TREE(TREE(3)), mapping each of the natural numbers to its own characteristic set. Mathematicians like Jon von Neumann and Ernst Zermelo saw the foundations of numbers and arithmetic lurking deep inside the dynamics of sets. Zero had transformed into the empty set – the set of nothing. It was the seed from which we grew the tree of all natural numbers.

We can find zero in this marvellous abstraction, but does it really exist? There is no consensus here. The Platonists argue that zero exists, along with all the other numbers, but only in an abstract sense, outside of space and time. The nominalists take a more practical view. They believe that numbers only exist to count the things we see in the real world – the loaves, the fish, the jars of oil – so they deny the existence of an emancipated number. The fictionalists deny the existence of numbers altogether! As for me, I believe in numbers. I see zero in the abstraction of the empty set, and in the empty set I see symmetry.

Why? Let me explain with nothing.

We need to distinguish nothing from Nothing. Nothing – with a capital N – is an absolute notion and far more difficult to comprehend. We shouldn't think of it as something we can create by taking things away – by removing apples or oranges or molecules of air or even the laws of physics. We can create a vacuum, but we can never create a Nothing. A true Nothing cannot be obtained from something or have the potential to be something. You cannot do anything to it. If it exists – and it's hard to see how it can – we must be disconnected from it.

But this isn't what we are really interested in here. We are interested in the weaker form of nothing, with a small n. This is not disconnected from us – we can reach it by taking things away, and that's how we connect it to the symmetry of zero. For example, when you have a pile of apples, you can subtract them away until you have zero apples. You could do the same with the oranges or molecules of air or even dinosaur bones. This weaker form of nothing is relative, not absolute. However, the important point for us is that zero apples and zero oranges are indistinguishable from one another. Each of them is identical to the empty set – to nothing. In a way, you could say that zero – or nothing – is something that remains unchanged even if you change the units: zero apples, zero oranges, zero dinosaur bones – we cannot tell the difference between them. Under zero, all things become equal. In other words, zero is symmetry – the symmetry of nothing.

This connection between zero and symmetry is more than just mathematics and philosophy. It is woven into the fabric of the universe, underpinning its physical laws, ordering the push and pull of fundamental particles. As we'll see in a moment, it's the reason that energy is neither created nor destroyed, or that light will travel at the speed of light. Perhaps the greatest discovery of the twentieth century is that ours is a universe that is filled with a huge amount of symmetry. It's a universe that is filled with zero.

FINDING ZERO

When the British government announced a nationwide lockdown to control the spread of coronavirus in the spring of 2020, my wife and I took it in turns to home-school our two girls. More often than not, we ignored the school's lesson plan and free-styled. My wife taught them to make a home biosphere so they could learn about ecosystems, while I would help them code silly computer games on *Scratch*. Of course, we couldn't stray too far from the curriculum and occasionally we'd run through the material sent by the teachers. It was on one such occasion that my younger daughter and I began to study symmetries.

She was given various shapes and asked to identify the lines of reflection. For example, for a square she had to draw the lines through the diagonals and the centres of opposite sides. I decided to ask her if there were any other symmetries she could see. Her class had only really been introduced to reflection symmetry so, at first, she was hesitant. With some gentle prompts she began to rotate the square about its centre, and after a quarter of a turn (ninety degrees) she realized the square looked exactly as it did before. We played the same game with pentagons, rotating through a fifth of a turn (72 degrees) and hexagons, through a sixth of a turn (60 degrees). At this point my artistic skills began to break down, but she had already understood. All these shapes possessed a special rotational symmetry depending on the angle of rotation. This and the reflections are examples of *discrete* symmetries, non-trivial jumps that leave something unchanged.

That something can be nature itself. To unpick the discrete symmetries of nature we need to peer deep into its microscopic kingdom and look for the corresponding zeroes. One possible symmetry involves trading all particles for their antiparticles, and vice versa. Is this symmetry present in nature? Well, if it were, there would have to be a zero – it would be the difference between the number of particles and antiparticles in our universe. But this difference is not zero – we see around 10^{80} particles and only a handful of antiparticles. That is an enormous stroke of luck. An equal number of particles and antiparticles would have annihilated each other into oblivion moments after the Big Bang, leaving behind a bath of radiation and a dead universe. We still don't know how or why the universe gave us this lucky break – why it broke the suicidal symmetry between matter and antimatter.

After my daughter and I had discussed the discrete symmetries of squares and hexagons at lockdown symmetry school, I drew a circle. I asked: how far do you have to rotate the circle to leave it unchanged? The answer, of course, is any angle you want. We are no longer restricted to multiples of 90, 72 or 60 degrees, as we were for the other shapes. You can rotate a circle continuously, through any angle about its centre, and it will always look exactly the same. This means we have a *continuous* symmetry, rather than a discrete one. In nature, continuous symmetries are responsible for some of the most important principles in physics.

For example, the laws of physics that Newton contemplated almost four centuries ago still apply today. They will apply in another four hundred years, or even a thousand years, even if they are only contemplated by the computer-generated scientists of the future. Although nature is free to evolve with the passing of time, the fundamental laws of physics are thought to remain the same. That is a continuous symmetry. The corresponding zero can be found in the blood-letting epiphanies of Julius von Mayer.

You may remember him from Chapter 'A Googol'. Von Mayer was the ship's physician who studied the colour of sailors' blood in the warmth of the tropical sun, stumbling on the fact that energy cannot be created or destroyed – it is always conserved. But *why* is energy conserved? It doesn't just happen by accident or divine authority – it follows from the fact that the laws of physics remain the same even as you travel through time. Conservation of energy follows from the continuous symmetry of time.

To get an intuitive feel for why this is true, let's think about what would happen if things were different and the laws of physics *changed* over time. For example, what if gravity became stronger overnight? Well, now it is easy to create energy from nothing. All you have to do is pick this book up off the floor, place it neatly on a shelf and leave it there overnight. In lifting the book you do work, passing on some energy which it then stores as gravitational potential energy. The next morning, when you're feeling a little heavy, the book is storing more potential energy because gravity has grown stronger. If you allow it to fall back on to the floor, it will release that energy, equivalent to more than you put in the day before. Well done – you have created energy from nothing, all thanks to the laws of physics changing over time. In contrast, in our universe, the laws of physics always appear to stay the same, so energy is never created or destroyed. It is always conserved.

Whenever you have a continuous symmetry, there is a corresponding conservation law. Here is another example: the fundamental laws of physics are thought to stay the same as you move around in space. They are the same in your house as your neighbour's house, or even the house of an alien on the shores of Sagittarius. This symmetry leads directly to the conservation of momentum. In a similar vein, the fact that the laws of physics are the same for a rotated universe leads to the conservation of angular momentum. With each of these and other continuous symmetries, we find a corresponding zero. It's just the total change in energy, momentum, angular momentum or some other conserved quantity.

This profound connection between symmetry, conservation and zero was discovered by the savant of symmetry, Emmy Noether. Einstein described Noether as a 'mathematical genius', while others put her scholarly credentials on a par with those of Madame Curie. Despite her considerable talent, she spent her entire life battling the prejudice of those around her. First, they took issue with the fact she was a woman and, later, that she was a Jew. Noether grew up in an academic German family in the late nineteenth century. Girls like her, from respectable middle-class families, attended finishing schools and were expected to pursue artistic interests. But Emmy resisted and began attending lectures in mathematics and languages at the University of Erlangen, where her father was a professor. As a woman, she wasn't allowed to enrol properly as a student. Her attendance at lectures was at the discretion of the lecturer and she was allowed to participate only as an auditor. Noether was one of just two women studying at Erlangen at the time. There were nearly a thousand men.

Even after she was awarded a PhD and began teaching at the Mathematical Institute, she did so as a second-class member of staff, without title or pay. But her brilliance was starting to attract attention. David Hilbert and Felix Klein fought hard to bring her to the University of Göttingen. They faced resistance, with many of their colleagues asking questions to the effect of: 'What will our soldiers think when they return to the university and find that they are required to learn at the feet of a woman?' But Hilbert and Klein prevailed and Noether moved to Göttingen in 1915. Of course, she wasn't paid and her lectures could only be advertised under Hilbert's name. It was in Göttingen that she began to see the interplay between symmetry and the conservation laws of nature. In keeping with her lowly rank, she wasn't allowed to present the work to the Royal Society of Sciences. Felix Klein presented it on her behalf.

When the Great War was over, German society began to slowly change, especially for women, and at the beginning of the 1920s Noether began to collect a small salary for her work at the university. Although she was receiving greater recognition away from Göttingen, she was never elected to the Academy of Science or even promoted to full professor. A decade after receiving her first paycheck, she would be expelled from her post, along with other Jewish and 'politically suspect' academics, as the Nazis took control of Germany. Noether fled to America, taking up posts at Bryn Mawr College in Pennsylvania and at Princeton. She died of a tumour two years after arriving in the United States. It wasn't the only tragedy to befall the Noether family. Emmy's younger brother, Fritz, also fled the Nazis, taking up a mathematics professorship at Tomsk State University in the Soviet Union. A few years later he would be imprisoned, accused of anti-Soviet propaganda and executed.

Emmy Noether's ideas have dominated fundamental physics in the twentieth century, as the quest to understand nature has become a quest to understand its symmetries and conservation laws. You can see a really important example of this by rubbing a piece of glass against a polyester jumper. No doubt you will generate some static, as electrons are torn off the glass and deposited on to the jumper. The glass now carries some positive electric charge, while the jumper carries a negative one. The two, however, are in perfect balance, so the total charge is still zero. This is thanks to the fact that electric charge cannot be created or destroyed. According to Noether, this conservation law must come from a continuous symmetry. So, what is that symmetry? It turns out that the theory of charged particles, like electrons and positrons, comes along with an internal dial. It's nothing more than a label that tells us the language we need to use to talk about what charged particles are doing. The language isn't English or Spanish but the mathematical language of complex spinors, whatever that might be. We're not going to get into the detail. All we need to know is that as the dial turns, the spinors turn too and in such a way that the physics doesn't change. In the end, it is the continuous symmetry of this internal dial that guarantees charge conservation.

The symmetry of electromagnetism is actually much more powerful than what we've just described. To see why, we need to put the universe in a box and think about ways in which you might conserve charge. For example, is it possible for a charged particle to disappear from right in front of your nose, only to reappear *instantly* on the other side of the road? Bizarre as it sounds, if all we cared about was conserving charge, this should be absolutely fine. After all, the charged particle jumped in an instant – it never actually left the universe. However, the moment we invoke Einstein and the spirit of chapter '1.0000000000000858', we know that charged particles cannot jump through space at infinite speed, faster than the speed of light. To be consistent with relativity, it turns out that charge must be conserved at a *local* level, at each and every point in space and time. Said another way, the overall charge in front of your nose, or anywhere else in space, cannot change in an

instant. This promotes the corresponding symmetry to a *local* symmetry. We should no longer be talking about a single dial for the whole
universe, but an infinite number of dials peppered across every point in
space and time, pointing any which way but loose.

This souped-up localized version of the symmetry is known as a
gauge symmetry. To understand its implications, I want you to imagine
my street as the universe, with every house corresponding to a point
in space. In our house there is me, my wife and our two daughters;
our neighbours to the left are Gary and Lynne, to the right Pete and
Steph, Ljupcho and Lilia a bit further along, Ian and Sue across the
road, and so on. It's all very sociable and people are often seen chatting over garden fences.

Suppose that each house contains a language dial. Right now, they
are set to 'English' everywhere, so everyone is speaking English. This
makes communication easy. If my wife decides to organize a party, she
can tell Steph, in English, who will then tell Lilia, also in English, and
so on. The message quickly gets around. But what if all the dials begin
to move, continuously, from English to American English, on to other
languages, until they all settle on French? Now everyone is speaking
French, but is that a problem? Of course not. If my wife wants to
organize another party she can tell Steph, in French, who will then tell
Lilia, in French, and so on. Once again, the message gets around. In
fact, you might even say that the message is conserved, thanks to the
symmetry of the dial.

But we also said that the symmetry was better than that. It was
souped-up and localized. That means the different dials don't need to
turn in unison. It could be that our dial moves to French, Gary and
Lynne's to German, while Pete and Steph end up speaking Swahili. In
the end, everyone on the road could be speaking a different language –
does this mean my wife will struggle to organize yet another party?
No: nature finds a clever way to adapt – *gauge theory*. It provides
every home with a bespoke language dictionary to help them connect
with their immediate neighbours. Ours contains translations from
French into German so that we can connect with Gary and Lynne, or
into Swahili so that we can connect with Pete and Steph. The message
about the party can still get around. Everyone everywhere in the universe of our street is allowed to set their dial to whatever language

they choose because nature provides them all with the right dictionary. Physicists like to call these dictionaries *connections* or *gauge fields*. They help to carry the messages to and fro. That's why we think of gauge fields as the forces of nature. In electromagnetism, the gauge field is the electromagnetic field, and the corresponding quanta is the photon – a particle of light. It helps to carry electromagnetic messages between charged particles.

Now that we have this fancy new souped-up symmetry, where is the zero? It turns out it's lurking inside the dictionaries. One thing we can ask is how much energy it costs to wiggle the gauge field/dictionary or to change it in some way. After all, the harder it is to wiggle, the heavier it must be. Think about shaking the tails of a mouse and an elephant with the same force. The elephant will wiggle far less because it is so much heavier. There is a sense in which the same is true of the gauge field – if we can change it at very little energy cost, we know it's very light, and if we can't, we know it's heavy. So, which is it? The answer lies in the gauge symmetry. What happens if my neighbours decide to reset their dial and change to yet another new language? We know this isn't a problem. Thanks to the symmetry, they should be able to do this without any physical consequences – there should be no energy cost. What does happen, of course, is that nature adapts to the changes by updating our dictionaries. In other words, there must be ways in which you can change the gauge field for free, without any energy cost. This means the field is as light as it can possibly be. It is *massless*. That's the zero – the mass of the gauge field and its corresponding quanta. Thanks to the gauge symmetry of electromagnetism, the photon has vanishing mass, forcing it to travel at the speed of light.

It seems that nature has a real appetite for symmetry and especially gauge symmetry. Gauge symmetries give you forces. They are at the heart of our understanding of gravity, the strong and weak nuclear forces and, of course, electromagnetism. The idea has dominated physics for almost a century. As we probe deeper and deeper into the microscopic dance of subatomic particles with ever more powerful particle accelerators, we see more and more symmetry. The closer we look, the more beautiful – the more symmetric – Nature becomes. And with every new symmetry, there is a zero.

When the ancient Babylonians wrote down the first zero, they did so in the interests of good accounting, to keep better records of food, livestock, people and goods. But zero was a number with too much personality, always destined for danger and excitement. In time, it would dance with the Devil, becoming one with the void and the absence of God. It is strange to think that a number condemned as heresy for so long should exist at the very core of what nature really is. In mathematics, zero is the empty set, an avatar of symmetry that can also be found in the physical world. Our universe is filled with zeroes, signs of symmetry in the clockwork of fundamental physics, from the vanishing mass of a photon to the vanishing changes in charge and energy.

As we will see in the next two chapters, nature contains some other little numbers: numbers that are much smaller than one but not quite zero. An example of this is in the mass of the electron – it doesn't vanish, but it is far smaller than the mass of all the other heavy particles, like the quarks, or the Higgs boson. This betrays a symmetry, albeit one with a slight imperfection, like a blemish on the face of perfect beauty. But there are also the little numbers that are still not understood, that betray no known symmetry. These are the mysteries of an unexpected world, of fundamental particles that should have stayed hidden, and of a universe in which you and I should never have been born.

0.0000000000000001

THE UNEXPECTED HIGGS BOSON

It was 4 July 2012. Across the United States, families were celebrating Independence Day, but the real excitement was in a lecture theatre close to the foothills of Mont Blanc, near the Swiss–French border. The theatre was the biggest hall at CERN, the European organization for nuclear research, home to the largest and most technologically advanced experiment in history. Here, scientists had built a Big Bang machine, a circular collider that accelerated subatomic particles almost to the speed of light – and then smashed them together. Their goal was to squeeze huge amounts of energy into tiny regions of space, but with just enough control to record what happened, to peer into the clockwork of fundamental physics. In the summer of 2012, they had seen something important in the aftermath of some of their collisions – and they were ready to tell the world.

Gathered in the audience that day were five giants of physics: Tom Kibble, Gerry Guralnik, Carl Hagen, François Englert and, of course, Peter Higgs. Along with their friend and colleague Robert Brout, who had died a year earlier, these were the 'gang of six' who had been instrumental in understanding the origin of mass in a world dominated by symmetry. Although widely accepted by this time, their theory had not yet received experimental confirmation, a prerequisite for a Nobel Prize and the holy grail for any theorist. All that changed on American Independence Day, when the team from CERN announced its results to these five wise men and half a million more watching via the internet. They had discovered a new particle whose mass was around 125 GeV – and they were pretty damn sure it was the Higgs boson.

There was much to celebrate, a triumph for both theory and experiment. With the might of its particle collisions, CERN had re-created the soldering furnace of an infant universe, a primordial crush of quarks, gluons and other cosmic ingredients. However, hidden among the celebrations on the morning of 4 July 2012 was a dark secret, something unsettling, something to worry all the theorists gathered in the audience. It was hidden in the following sentence:

> *They had discovered a new particle whose mass was around 125 GeV . . .*

125 GeV. In plain English that's a mass of around 2.2×10^{-25} kilograms, once we have converted the units.[1] That is almost a billion billion times less than the mass of a fairyfly, a tiny wasp that happens to be the smallest insect in the world. Of course, we wouldn't want to compare a fairyfly made up of billions and billions of atoms to a single Higgs boson but, even so, the Higgs boson is far lighter than expected. By all accounts, it should really be a particle heavyweight, much heavier than an electron or a proton. It should really weigh a few micrograms. As it happens, that's about the same as a fairyfly.

I know what you're thinking: what do fairyflies have to do with the Higgs boson? The answer is nothing – not directly anyway. It turns out that a fairyfly weighs almost as much as a quantum black hole, the smallest and most compact object gravity will allow. The insect and the quantum black hole might have more or less the same mass, but the black hole packs it all into a space that is more than a million trillion trillion times smaller. In fact, it squeezes eleven micrograms into a ball of radius the Planck length, around 1.6×10^{-35} metres. That's the scale at which gravity starts to break down the fabric of space and time. It's an unimaginably small distance but one that should be really important to the Higgs. If we push our understanding of physics all the way down to this tiny threshold, the Higgs should be dragged along by the bubbling world of quantum mechanics so that eventually it goes toe to toe with quantum gravity. I'll explain the details later on in this chapter. For now, try and accept that the Higgs should weigh as much as a fairyfly, and almost as much as a quantum black hole, but that it doesn't. It is 0.000000000000000 1 times lighter and no one knows why.

In the previous chapter, I tried to convince you that tiny numbers demand an explanation. When you encounter a zero, nature is taunting you with its beauty – its symmetry. After all, there is perfection in zero. But what about a small and non-zero number – a number like 0.000000000000000 1? That is close to perfection but not quite there. It is symmetry with a slight blemish, like a face perfectly balanced from left to right save for an extra tiny freckle on the left cheek. In the physical world, unless you are under the spell of symmetry, you don't expect to see numbers that are big or small. The ratios you see should be unremarkable, roughly around one or some other single digit number. If you do see some remarkable numbers, then chances are something remarkable is going on.

There is a little experiment you can do to convince yourself of this. Ask ten of your friends to randomly pick an irrational number between −1 and 1. Remember, an irrational number is one that cannot be written as a fraction, so your friends might choose numbers such as

$$\frac{1}{\sqrt{2}}, \frac{\pi^2}{18} \text{ or } -\frac{1}{\sqrt{13}}.$$

When they have finished, add all their numbers together and drop the overall sign. What do you get? If you get something that is smaller than 0.000000000000000 1, then it is certainly remarkable. Your friends have somehow conjured up a very unlikely cancellation. Without some sort of conspiracy, this just isn't going to happen. The answer you will get will not be close to zero. It'll just be a number. Nothing too big, too small or especially fancy.

We can use this philosophy to help select our best scientific models. To see how it might work, we should return to the early part of the sixteenth century, when most people believed the Earth to be at the centre of the whole universe. Astronomical observations at the time were not inconsistent with this view. They could be explained by the ancient model of Ptolemy of Alexandria: of equants and epicycles, with planets moving on circles that themselves moved on other circles. The details don't really matter, save for the fact that the Earth was assumed to be still while all the other planets were moving at comparable speeds. In 1543, this view was challenged by Nicolaus Copernicus. Born in the

Kingdom of Poland, Copernicus was a canon of the Catholic Church, with a keen interest in mathematics and astronomy. He was inspired by the writings of Cicero and Plutarch, and argued that the Earth should no longer be fixed – it should be free to move alongside all the other planets. He then proposed a heliocentric model, with the Sun at the centre of the universe and the Earth in deferential orbit. At the time, the astronomical data wasn't precise enough to prove or disprove this radical new idea, so most philosophers went with their gut. The Copernican model seemed to challenge common sense or, worse, the Bible. Copernicus himself had anticipated this response. Fearing the inevitable scorn, he held on to his work for decades, delaying its publication until the very last moments of his life.

Copernicus's contemporaries might have taken another, more enlightened, view based on unremarkable numbers. In a heliocentric model, with the planets orbiting around the Sun, they are all moving at roughly the same speed. Mercury is fastest, cruising at around 107,000 mph, Venus next at 78,000 mph, then Earth at around 67,000 mph, Mars at 54,000 mph, and so on. Although the planets clearly slow down as they move away from the Sun, the ratio of their speeds is always unremarkable – never too big, too small or too special. This is certainly not the case in Ptolemy's geocentric model. Because the Earth is assumed to be still, in contrast to all the other planets, the ratio of its speed to any other planet is vanishing. Therefore, the geocentric model contains a zero – a remarkably small number – and nature doesn't tend to do remarkable numbers without a very good reason. Ptolemy's supporters should have asked about this zero. Why should the Earth stay still? In the heliocentric model, we can justify the stillness of the Sun on the grounds that it is so much more massive than the planets and has far more inertia. But the inertia of the Earth is roughly the same as that of Venus or Mars. There is no good reason to assume the Earth is still and we cannot justify Ptolemy's zero. Even if Ptolemy and Copernicus could not be separated using astronomical data, we could have argued in favour of Copernicus. After all, his model was a good enough fit to observation and did not rely on any remarkable numbers that could not be explained.

This criteria for selecting theories is known as *naturalness*. A theory is natural if it doesn't contain any unexplained and finely tuned inputs.

You can have some small or precise numbers, but only if you understand the physics that underpins them. Without that understanding, the likelihood is that something is missing, or the theory is fundamentally wrong, as in the case of geocentric cosmology. Of course, to some extent, naturalness is just an aesthetic consideration – it should never be used ahead of experimental data. But when the data isn't quite ready to guide us, it seems that naturalness is an able deputy. Whenever we see a small number that we cannot explain or justify, we start to think hard about why it's really there. What is the symmetry? What is the new physics we are missing?

The case for naturalness is compelling, not just for mathematical reasons, but because we see it realized so often in nature. For example, at the end of the previous chapter, we learned how the photon has vanishing mass. This wasn't just a randomly chosen zero. It was there thanks to the gauge symmetry of electromagnetism – the freedom to set an internal dial at each and every point in space. There is also a zero lurking within nuclear physics, wrapped up inside the internal structure of protons and neutrons. The fundamental quarks that make the protons and neutrons are held together by gluons. These also have vanishing mass, thanks to another gauge symmetry, this time associated with the strong nuclear force, as opposed to electromagnetism.

But naturalness isn't just about zero. It's also about the surprisingly small. The electron isn't massless, like the photon or the gluon, but it's at least a million times lighter than we might naively have expected. That small number – a factor of a millionth or less – demands an explanation. And we have one. The electron is light because of a symmetry. It's not a true symmetry, as that would render the electron massless. Instead, it's an *approximate* symmetry. We aren't going to worry too much about what the symmetry is, but about what it does: it stops the electron from getting too heavy. This is a very good thing. If the electron had been just three times heavier, it would have destabilized the hydrogen atom. There would have been no such thing as chemistry or biology and you and I wouldn't ever have existed.

Perhaps the greatest victory for naturalness came in the so-called November Revolution of 1974, when teams at Stanford Linear Accelerator Center and Brookhaven National Laboratory began seeing evidence for a new species of quark, known as charm. Just a few months earlier, at Fermilab near Chicago, two young theorists, Mary Galliard and Benjamin Lee, had been studying the difference in mass of two versions of a high-energy particle known as a kaon. They realized that without some new physics just around the corner, naturalness would fail. They guessed the new physics might take the form of a new species of quark, and the charm showed up right on cue, just where naturalness said it should be.

. Fast-forward almost forty years to the gathering at CERN on American Independence Day in 2012. The Higgs had arrived, joining up the dots of fundamental physics, explaining how a universe had hidden so much of its underlying symmetry. But as we have seen, there was something unnatural about the whole charade. The Higgs was a billion billion times too light. The theory we were celebrating with such gusto contained a tiny number, perhaps as small as 0.0000000000000001. Nature doesn't do small numbers, not without a good reason. So why was the number there? What was the new physics that would save us from it? What was the new symmetry?

For Galliard and Lee in the summer of 1974, the new physics was waiting just around the corner and naturalness was saved. But a decade on from that gathering in CERN in 2012, we are still waiting to understand why the Higgs is teasing us with such a tiny number. The new physics that naturalness has promised still hasn't shown itself. Has naturalness finally failed? Are we doomed to live in an unexpected and unlikely universe without ever understanding why? We need to take a closer look at this troubling new particle. In fact, we need to take a closer look at all the particles.

A QUICK REFERENCE GUIDE TO ALL THE PARTICLES YOU'LL ENCOUNTER IN THIS CHAPTER

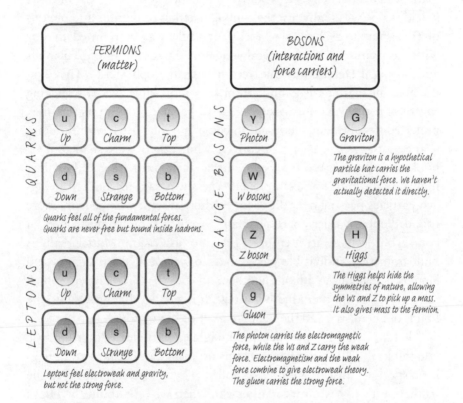

FERMIONS (matter)

QUARKS

| u Up | c Charm | t Top |
| d Down | s Strange | b Bottom |

Quarks feel all of the fundamental forces. Quarks are never free but bound inside hadrons.

LEPTONS

| u Up | c Charm | t Top |
| d Down | s Strange | b Bottom |

Leptons feel electroweak and gravity, but not the strong force.

BOSONS (interactions and force carriers)

GAUGE BOSONS

γ Photon

W W bosons

Z Z boson

g Gluon

G Graviton

The graviton is a hypothetical particle that carries the gravitational force. We haven't actually detected it directly.

H Higgs

The Higgs helps hide the symmetries of nature, allowing the Ws and Z to pick up a mass. It also gives mass to the fermion.

The photon carries the electromagnetic force, while the Ws and Z carry the weak force. Electromagnetism and the weak force combine to give electroweak theory. The gluon carries the strong force.

PARTICLE PARTICULARS

Aristotle would have hated the Higgs boson. In fact, he would have loathed all the particles. He would have recoiled at the idea that nature's kaleidoscope was really just the coming together of billions and billions of these miniature building blocks. Aristotle was committed to war with the atomists – to campaigning against the teachings of Leucippus and his pupil Democritus, the very first particle physicists. They had declared all matter to be built from tiny, indivisible pieces frolicking together in the vacuum of space. These particles, or 'atoms' as they preferred to call them, were said to come in a huge variety of different shapes: some were concave or convex, while others were shaped like a hook or even an eye. They believed that their particles could explain human sensations. For example, bitterness was said to come from jagged particles passing over the tongue, while sweetness came from more rounded ones. Modern particle theory is a little more sophisticated, of course, but at its heart it embraces the atomist vision. Matter really is built from tiny, indivisible pieces, but now we call them quarks and leptons. They dance with one another and with the force carriers, which are themselves another kind of particle. It is a ballet that grows into the bonds of chemistry and the life-giving art of biology.

What do you think of when you imagine a particle? I don't suppose you think of a hook or an eye, as the atomists did. Perhaps you imagine a speck of dust or a glitter of pollen. That's certainly closer to the truth, but it's not what we really mean when we talk about the Higgs or the electron or any of the other fundamental particles. To appreciate what a particle really is, we first need to talk about fields. When I was a kid, I just thought of a field as something I could play football on, but in physics there are other kinds of field, invisible forces that push and pull. There are electromagnetic fields, wielding this invisible power through the tug of a magnet or the violence of a lightning storm. There are gravitational fields, controlling the motion of planets or tearing apart stars whenever they drift too close to a black hole. But you can also have electron fields, quark fields and even a Higgs field. There isn't really anything fancy or mysterious about a field. It's just something you can map across space and time, taking on different values at different points. For example, you can talk about the temperature field on a weather map labelling the inevitable cold in England and the warm in Italy or Spain. You can also talk about the field of atmospheric pressure mapping out the pressure of the air, or the density field in a galaxy mapping out the distribution of interstellar gas or clumpier objects like stars and planets. The electromagnetic field is just another one of these maps, a bunch of numbers labelling every point in space and time, only now it encodes the strength of the electromagnetic background.

Of course, the electromagnetic field is superior to the others in one sense: it is an example of *a fundamental field* – it cannot be unpicked to reveal any underlying structure. There are other fundamental fields such as the electron field, the Higgs field, fields for the up quark, the down quark, the Z boson and, of course, the gravitational field. The list goes on. Some of these fields, like the electron field, only make sense in the quantum realm – as quantum fields – while others, like electromagnetism and gravity, can survive on macroscopic scales. We'll explain how this works in a moment. But whatever the field, we should just think of it as a bespoke map, a series of numbers spread throughout space and time, encoding the relevant physical effects. For example, if the electron field is vanishing everywhere, you can be sure you won't find any electrons.

Where do the particles appear in all of this? As we saw in the chapter 'Graham's Number', a particle is really just a tiny vibration – it's a quantum ripple in a quantum field. Think of the surface of the sea as analogous to the value of some fundamental field, its level slowly moving up and down with the ocean swell. On top of this swell you could imagine a tiny ripple – that would be equivalent to a particle. The ripples in different fields give different particles. A ripple in the electron field gives an electron, a ripple in the electromagnetic field a photon, in the gravitational field a graviton, in the up-quark field an up quark. We could go on.

Particles also tend to be described as either *real* or *virtual* – you can have real photons, but you can also have virtual ones. The same is true of electrons, quarks, gluons and all the other fundamental particles. It all sounds a bit more mysterious than it actually is. A real particle is one you can hold in your hand, like a real photon being emitted by candlelight or a real electron fired through a pair of slits in the classic experiment of quantum mechanics. You cannot hold a virtual particle. This isn't because it's lost in the aether of some virtual reality game. It's because it's not really a particle at all. It's just a disturbance in a field caused by other particles and other fields. For example, an electron will create a disturbance in the electromagnetic field that is felt by a passing electron, and vice versa. This disturbance is what pushes the electrons apart. You can even think of it as a ripple – a photon – but it's not a true particle in any meaningful sense. It's a virtual one. The ripples of a virtual photon don't automatically move at the speed of light like they do for real photons, nor is there any way to intercept them.

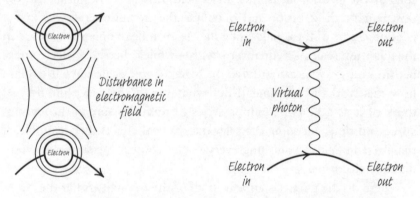

The two electrons cause a disturbance, or ripple, in the electromagnetic field – that's what we mean by a virtual photon. On the left we have the more physical picture showing the contours in the electromagnetic field; on the right we have the kind of diagram a particle physicist would draw to mean exactly the same thing. The latter is an example of a so-called Feynman diagram, named, of course, after Richard Feynman.

Virtual particles are just a convenient way to think about how different fields can influence one another. The analogy you often hear is of two ice skaters throwing a ball to one another. Inevitably, when they throw the ball, or catch it, they are pushed back a little, as if being repelled by the other skater. The skaters are like the electrons feeling the electromagnetic repulsion while the ball is like the virtual photon carrying that effect from one skater to the other. The analogy doesn't work so well for an attractive force, but we still think of them as virtual particles passing between charged objects.

Most particles also come equipped with an intrinsic ability to 'spin'. We've had hints of this ever since two Germans, Otto Stern and Walter Gerlach, began playing around with magnets and silver atoms in the early 1920s. Spin is really a form of angular momentum, the kind of momentum you normally associate with rotational motion, like the spin on a ping-pong ball or the waltzers at a fair. While this is easy enough to imagine for a ping-pong ball – even a quantum ping-pong ball – it is a little trickier to picture what it really means for fundamental particles. This is because they are *infinitely* small. When an ice skater is pirouetting on the ice, they bring their arms in so that they spin more quickly. This works because their angular momentum is conserved. Angular momentum depends on two things: how fast you rotate and how spread out you are. When they tuck their arms in, they need to compensate for the loss of spread by spinning faster. For an infinitely small particle to register any angular momentum at all, it's as if it were spinning infinitely fast. That obviously can't be right, so what's really going on? With point-like particles, what we are talking about is an *intrinsic spin*, an ability to look and act as if they are spinning without actually whizzing around in an infinite frenzy. Think of it like a politician. Their job is to look and act as if they have your best interests at heart. Whether they actually do is quite another matter.

With this caveat, let's run with the image of a particle as a ping-pong ball, scaling it down to microscopic size. Particles of different spin behave differently when you rotate them. Suppose you draw a smiley face on top of the ping-pong ball. As you rotate the ball, your perspective changes steadily as the smiley face moves around. It is only after one entire revolution that things look exactly as they did at the start. This is what happens for the photon and other so-called 'spin one' particles. To take them back to their original quantum state, you must perform a complete revolution. To capture what happens for the graviton, which is 'spin two', we need to draw an identical smiley face on the opposite side of the ball. Now as we rotate the ball, the original picture is restored twice in one revolution, after 180 degrees and again after 360. A spin-two particle will return to the same quantum state twice in a single revolution. A spin-three particle three times, and so on.

All the particles we have just described have integer spin, but you can also have particles of half-integer spin. What happens when we rotate a spin half particle? Well, this is where things get a little tricky. Instead of a ping-pong ball, I want you to consider the vampire squid, scaled down to quantum proportions. Now after one entire revolution, you probably expect the squid to look exactly as it did before. Except that it doesn't. The squid has turned itself inside out. It's *inverted* itself. As it happens, vampire squids can actually do this but, in the language of quantum mechanics, what we really mean is that the probability wave has flipped over – peaks have become troughs and troughs have become peaks. This always happens for particles of half-integer spin – after one entire revolution they switch from one state to an opposite state, as if they have been turned inside out! It is only after completing a *second* revolution that things get back to normal.

Spin allows us to put particles into two different camps. On the one hand, we have the particles of integer spin – the *bosons* – responsible for carrying all the forces of nature. The photon is a boson. It has spin one and carries the electromagnetic force. We also have the W and Z bosons and gluons, spin-one particles that carry the forces of nuclear physics. And then there is the graviton – the undetected quanta of spin two said to be responsible for carrying gravity. Light particles like the photon carry their force over very large distances. However, when a heavy particle is carrying a force, it runs out of steam more quickly and the range of the force is shorter. That's what we see for the W and Z bosons that carry the weak nuclear force.

What about the particles of half-integer spin, like electrons and quarks? These are the *fermions*. They are responsible for giving the universe its guts. This is because fermions make matter. They make up all the solid stuff like stars, planets and a stick of Blackpool rock. There is a very good reason for this. Fermions do not like to be piled up in the same place doing exactly the same thing. In fact, in any quantum system it is completely forbidden for any two of them to be in the same quantum state. This is known as the Pauli Exclusion Principle, named after the brilliant German physicist Wolfgang Pauli, who we'll encounter much more in the next couple of chapters. It works like this: imagine two fermions floating around inside a cup of tea. What happens if you swap them around? Well, fermions are awkward little things. If you swap them around, they flip over the probability wave that describes the tea – a positive peak becomes a negative trough, and vice versa. It's the drama of the vampire squid turning itself inside out all over again. Now if the two fermions happen to be identical, your tea is in trouble. By identical, I mean genuine doppelgängers right down to their quantum DNA – same spin, same energy, same opinion about Brexit, and so on. If you swap them over, nothing will actually change. How can it? After all, they are doppelgängers. And yet we just said that everything is flipped over. If flipping a wave also leaves it *unchanged*, there can't have been any peaks or troughs to begin with! Wherever you look, the wave should be perfectly flat, sheepishly settled at zero. As this is really a probability wave, that means there is always a vanishing probability. In other words, a cup of tea with identical fermions has no chance. It cannot exist. As for the vampire squid, it will usually turn itself inside out to ward off predators. But if it just looked exactly the same inside as out, this strategy would fail and such a beast could never survive. That's Pauli's Exclusion Principle.[2]

Pauli was a colourful yet uncompromising scientist. Throughout his career, he was famously a perfectionist, a tenacious bulldog billed as the 'conscience of physics' for his feared ability to cast devastating judgements on the work of his contemporaries. Rudolf Peierls, once Pauli's assistant, reminisces about some of these criticisms in his memoirs. On one occasion, Pauli was asked for his opinion on a paper by a young and inexperienced physicist. He knew the work wasn't right but because he was so unimpressed with the incoherence of the argument, he declared it to be 'not even wrong'. The phrase has since become embedded in the lexicon of theoretical physics as a way to describe bad science. To be fair to Pauli, he could be just as cruel to his more famous colleagues. After he had spent a long afternoon arguing with the great Russian physicist Lev Landau, Landau asked if he thought everything he had said was nonsense. 'Oh no! Far from it. Far from it,' remarked the German, 'What you said was so confused that one could not tell whether it was nonsense or not.'

There is no exclusion principle for bosons. They are a sociable bunch, more than happy to pile up alongside each other in the same quantum state. In fact, it is this gregarious quality that often allows them to build up into gigantic beasts of macroscopic scale. This is particularly important for Bond villains who like to build enormous lasers that threaten mankind. A laser is a vast collection of real photons, many of which are in the same quantum state, their phases held together in lockstep. The macroscopic waves we see in electromagnetism and gravity are in fact just real photons and real gravitons piled up on top of one another in enormous number, something you could only ever do with bosons.

Although most of us are familiar with electromagnetism and gravity, the other two forces are less well known, mainly because they only operate on the scale of nuclear physics, deep inside the core of an atom. As we will see in a moment, this is a world of quarks chained together by gluons, and changing into one another with the help of a W or a Z. It is a microscopic mayhem made possible by the inevitable Higgs boson, capable of unleashing terrific power, from the life-giving warmth of the Sun to the terror of a nuclear apocalypse. As I said at the beginning, this complex zoo of subatomic particles would never have appealed to Aristotle and his ancient followers. But his enemy, Democritus, and the rest of the atomists? I think they would have loved it.

THE INEVITABLE HIGGS BOSON

Let's descend into the atom.

Suddenly you are in a tiny solar system, with planetary electrons in orbit around a microscopic 'Sun' known as the nucleus. Of course, these atomic orbits are not controlled by gravity, as in the real solar system, but by electromagnetism. The electromagnetic force between a negatively charged electron and a positively charged nucleus is roughly 1,000 trillion trillion trillion times stronger than the gravitational force. The nucleus is made out of protons and neutrons: the protons give it the positive electric charge it needs to cling on to the electrons, whereas the neutrons are electrically neutral, as the name suggests. Depending on the element, you can sometimes find a large number of protons piled inside the nucleus. Although the nucleus of a hydrogen atom has only a single proton, a nucleus of gold has seventy-nine. This brings us to our first atomic mystery: positive electric charges are well known to repel one another, so how do seventy-nine protons gather together in such a tiny space? Something must be pulling the neutrons and protons together with enough force to overcome the electromagnetic repulsion. We know it's not gravity – that's far too weak. It must be something stronger.

The strong nuclear force

If protons and neutrons were all we had to worry about, the story of the strong force would have been relatively simple. However, in the decades that followed the Second World War, particle physics declared itself to be richer and stranger than anyone had ever imagined. Photographs began to capture the tracks of cosmic rays as they showered their way through the Earth's atmosphere. They revealed a slew of new and magnificent particles, many of which were dancing to the tune of the strong nuclear force. There were pions and kaons, eta and rho mesons, Lambda and Xi baryons, all members of a wider family of particles now known as hadrons. To many, the task of keeping up with all the new discoveries was exhausting. Pauli, who was never short of an opinion, is said to have complained, 'Had I foreseen that, I would have gone into botany.'

Pauli may have frowned at the zoo of noisy new discoveries, but a young man from Lower Manhattan named Murray Gell-Mann was starting to see patterns in them. Along with the Israeli Yuval Ne'eman, Gell-Mann examined the properties of these new particles and arranged them into beautiful eight- and ten-folded patterns that would not have been out of place at the Alhambra in Spain. There was no way such organized elegance happened by chance – there had to be some underlying structure. It was Gell-Mann, inevitably, who figured out what it was. So did George Zweig, a young Russian-American who had just completed his PhD at Caltech, under the supervision of Richard Feynman.

Gell-Mann called them *quarks*, Zweig called them *aces*, but they were one and the same thing. These were the bricks that built protons, neutrons, pions and all the other hadrons. We now know of up to six types of quark: the up, the down, the strange, the charm, the top and the bottom. They are all fermions, some heavier than others and carrying different amounts of electric charge as well as other quantum properties like isospin, charm and strangeness. If three quarks are bound together, you have what is known as a baryon, like the proton or the neutron. Mesons, like the pion, are made from two, rather than three combinations of quark. The different combinations account for different particle properties. For example, the proton is made from two ups and a down. Given that the up quark has a fractional electric charge of +2/3 and the down quark -1/3, it follows that the proton carries a total of one unit of positive charge. A neutron is made of two downs and an up, which is why it ends up neutral.

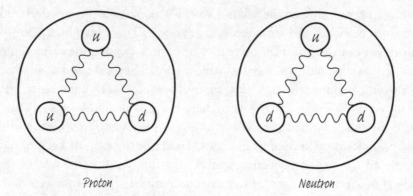

Proton   Neutron

The underlying structure of the proton and the neutron. These are built from up and down quarks of different colour, held together by gluons.

At this point, the ghost of Pauli should be whispering in your ear. Quarks are fermions. How can the proton contain two up quarks or, in other words, two identical fermions? Isn't that forbidden by the exclusion principle? This would certainly be true if the up quarks really were identical, but they aren't. Quarks can also come in different 'colours', either red, green or blue. In the proton, if one of the up quarks is red, the other one must be either green or blue. This colour has nothing to do with what we normally think of as colour – it's just a label for a new type of charge. Feynman was particularly unimpressed with the confusion, declaring that the 'idiot physicists' must have been 'unable to come up with any wonderful Greek words' for this fancy new ingredient.

Perhaps it was a dig at Gell-Mann. The two men, whose offices at Caltech were just a few doors apart, had a prickly relationship. Feynman would often poke fun at Gell-Mann's obsession with naming things. He once told an apocryphal tale about how Gell-Mann came to him one Friday, desperately worried about finding a good name for a new type of particle he'd been working on. Feynman irreverently responded by calling them 'quacks'. The following Monday, Gell-Mann came up to him excitedly. He'd found the perfect word while reading a line in *Finnegans Wake* by James Joyce: 'Three quarks for Muster Mark.'

So not quack, as Feynman had suggested, but 'quark'.

He may not have liked Gell-Mann, but there is no doubt that Feynman respected him enormously. In 2010, I had the privilege of attending a conference in honour of Gell-Mann's eightieth birthday. It was held in Singapore and was a star-studded event, at least for a physics fanboy like me. As well as Gell-Mann, there were three other Nobel Laureates in attendance: Gerard 't Hooft, who we encountered in Chapter 'TREE(3)', Gell-Mann's student Kenneth Wilson and the Chinese physicist Yang Chen-Ning, also known as Frank, a name he took in honour of the American polymath Benjamin Franklin. George Zweig was also there. And yet, even though he was surrounded by some of the sharpest minds in the recent history of physics, Gell-Mann stood out. He oozed a confidence and an intellect that I had never seen before and haven't seen since. I admit, I was a little star-struck. Gell-Mann was, at the time, the last of a golden generation of physicists. This was a man who had sparred with Feynman at Caltech, who was a Nobel Prize winner by the age of forty and could easily have won two or three more in the years that followed. His mental capacity was far beyond that of an ordinary human. By the age of nine, he had memorized the *Encyclopaedia Britannica*, and he was also a skilled linguist, fluent in at least thirteen languages.

Gell-Mann's quarks, or quacks, are the building blocks of all matter, along with another family of fermions known as leptons. Leptons include the electron and its heavier cousins, the muon and the tau, as well as the delightfully named neutrinos, which we'll come across in a moment, when we talk about the weak nuclear force. Although leptons and quarks have much in common, they are also very different, and in a very important way. The leptons are immune to the strong nuclear force. They cannot feel it at all. But the quarks are imprisoned by it. It ties them together, trapping them for ever inside hadrons. Unlike leptons, quarks can never be free. This is the curse of *confinement*. Confinement means you will never find a quark wandering alone through the cosmos. It will always be chained up alongside other quarks in the cell of a proton or a neutron or some other hadronic particle. Those chains are made from gluons, the particle of incarceration and the carrier of the strong force.

Gluons don't just keep quarks prisoner; they also imprison each other. They will pull on other gluons, as well as quarks, squeezing the lines of force, so that, in the end, confinement confines them all. That's why we don't get to see the strong force in our macroscopic lives. Even though the gluon is massless, confinement keeps the force packed away inside a nucleus. We still don't understand this process properly. The problem is the subject of a million-dollar prize being offered by the Clay Maths Institute, so if you can figure it all out, you'll become rich.

It was, of course, Gell-Mann and his collaborators who pieced together what we do know in the early 1970s. Because the quarks and gluons were said to have 'colour', the theory became known as quantum chromodynamics, or QCD for short. The seed had been planted decades earlier when one of the delegates from Singapore, Frank Yang, and his American collaborator, Robert Mills, cooked up a fancy version of electromagnetism now known as Yang–Mills theory. This new theory contained its own force carrier, a new gauge boson which you might think of as the more complex and misunderstood cousin to the photon. When Yang spoke about it at Princeton, Pauli had repeatedly asked him about the mass of this proposed new particle. To Pauli this question was crucial, as no such particle had ever been seen. Yang didn't know the answer and was so taken aback by the ferocity of Pauli's attack that he sat down in the middle of his own seminar and took a moment. It was deeply embarrassing, and Pauli kept quiet after that, although he did send Yang a note the next day declaring that he had made it impossible for the two of them to talk afterwards. We now know the answer to Pauli's question. Thanks to the symmetries involved, Yang's force carrier had no mass at all. Tweaking the symmetries a little, but not the mass, Gell-Mann identified this new particle with the gluon, the chain that binds together protons, neutrons and the nucleus of an atom. It was the carrier of the strong force.

The weak nuclear force

My friend Smarty used to joke that if he had three kids, in the interests of science he would conduct an experiment. He would name one of his children Fantastic, another one Brilliant and the last one Rubbish, and then sit back to see how they all turned out. As it happened, he did have three kids but, luckily, his wife stopped him from going through with his original plan. The story always reminds me of the *weak nuclear force* and its unfortunate place alongside those that sound far more impressive: gravity, strong, electromagnetic. The irony is that the weak force isn't even the weakest of the four fundamental forces. That dishonour goes to gravity, which is over a trillion trillion times weaker.

Of course, the weak force isn't as strong as the strong nuclear force or even electromagnetism, but don't let that put you off. It is the sunshine of the subatomic world. And I mean this literally – the weak force is responsible for giving the Sun its life-giving glow. When two hydrogen nuclei are squeezed together inside the solar core, there is a chance that one of the two protons will shapeshift into a neutron, allowing you to create a heavy-form hydrogen known as deuterium. This is the first step in the nuclear fusion process that allows the Sun to generate so much energy. As we will see in a moment, it is the weak force that allows protons and neutrons to shapeshift into one another. It is the force of radioactivity.

It all started with a puzzle, as is so often the case in physics. On the eve of the First World War, a young British physicist by the name of James Chadwick travelled to Berlin to work with Hans Geiger. Geiger had recently developed his famous Geiger counter, which Chadwick used to measure the spectrum of radiation emitted through a nuclear process known as beta decay. At the time, beta decay was thought to occur when a heavy atomic nucleus spat out an electron. Like everything in the quantum world, the energy of the nuclei before and after the decay was expected to take on very precise values. If energy were to be conserved, as everyone thought it should be, the same should be true of the electrons that made up the radiation. But it didn't work out that way. Chadwick noticed that the electrons were allowed any amount of energy – their distribution was continuous. Beta decay seemed to go against the idea that energy was neither created nor destroyed. The result sent physics into turmoil. Even the great Niels Bohr was ready to give up on energy conservation, to cast aside the breakthrough made long before by Julius von Mayer as he examined the blood of his sailors. For his part, Chadwick ended up in a civil detention camp after finding himself stuck in Germany when war finally broke out. To be fair to his German guards, they let him set up a laboratory and supplied him with radioactive toothpaste to carry on his experiments.

Chadwick's puzzle would be solved by another German. It came in the form of an extraordinary letter sent by Pauli to the participants of a conference held in Tübingen in December 1930. Pauli had been unable to attend in person because he was due to attend a ball in Zurich. Nevertheless, his virtual contribution guaranteed the conference its place in the history of physics. Never content with a dull introduction, Pauli began his letter, 'Dear Radioactive Ladies and Gentlemen'. He followed it with a remarkable guess. Pauli suggested that the problem of beta decay could be solved by tiny little neutrons. The idea was that they could be spat out as radiation alongside the electrons, carrying with them the missing energy in Chadwick's experiment. Pauli's neutrons were not the same as the ones known to be lurking alongside protons in the nucleus of an atom. Those particular neutrons would be discovered by Chadwick within a year or two and were much heavier than the particle Pauli was proposing. Pauli was thinking about what we now call a *neutrino* – something that was little, light and electrically neutral.

When Pauli gave a talk about his little neutrons at a conference in Brussels in 1933, it left a deep impression on the father of fermions, Enrico Fermi. Fermi returned to Rome determined to piece together the details of Pauli's idea. When a nucleus spat out an electron during beta decay he realized that it wasn't spitting out an electron that was already there. Something else was happening, something completely new. A neutron inside the nucleus was decaying through a new and unknown force – the force we now know as the weak force. The product of that decay was a proton, an electron and one of Pauli's neutrinos. Well, technically, it was an antineutrino, but let's not worry too much about that. You shouldn't think of the neutron as being built out of protons, electrons and neutrinos and then disintegrating. It literally changes into them, like a subatomic shapeshifter. Once the transformation is complete, the proton will increase the atomic number of the nucleus, nudging it one place up the periodic table, while the electron and the neutrino are fired out as radiation. Fermi's new force – the thing that is responsible for all this radioactive drama – acts over an infinitely small distance, as if it were being carried by an infinitely heavy particle. It's what we now call a contact force: the neutron kissing the new particles – the proton, the electron and the neutrino – in a single place at a single moment in time. When Fermi submitted his work to the journal *Nature*, it was rejected for being too far removed from physical reality. *Nature* has since admitted it was one of the greatest editorial blunders in its history. As for Fermi, he took the rejection badly and decided he needed some time away from theoretical physics. He focused, for a while, on experiments and by the end of the 1930s he was awarded a Nobel Prize. Fermi had developed a method for slowing down neutrons, making them more accurate as projectiles for splitting atomic nuclei. He realized the potential for extracting huge amounts of atomic energy and paved the way for nuclear power on an industrial scale.

It's hard to spot a neutrino. The problem is that they have barely any mass and no charge, so they don't tend to do very much. It's just as well, as there are currently about 100 trillion neutrinos passing through your body every second. Thanks to this ability to remain incognito, the neutrino would not be discovered experimentally until 1956, more than two decades after Pauli and Fermi's original proposal. When he received the telegram informing him of the discovery, Pauli knowingly replied, 'Everything comes to him who knows how to wait.'

Six months after the discovery of the neutrino, the world of physics was shaken by an even more remarkable experiment. The scientist in charge was Chien-Shiung Wu, or Madame Wu as she was more commonly known. Wu had grown up in Liuhe, China, close to the Yangtze estuary, the daughter of a teacher and an engineer who passionately encouraged her academic interests. It was a progressive environment that allowed her education to flourish. As she would later note in an interview with *Newsweek* magazine, 'In Chinese society, a woman is measured solely by her merit. Men encourage them to succeed, and they do not have to change their female characteristics in doing so.' But when she arrived in the United States to begin her doctorate in 1936, at the University of Michigan she found a very different perspective.

Female students were not allowed to enter the new student centre through the front entrance – they had to sneak in through the side. Wu was so appalled by this she decided to head for the West Coast, to Berkeley, where attitudes were more liberal. Even then, she still had to overcome the fact that scientists weren't meant to look like she did. Wu was pretty and petite – the *Oakland Tribune* reported that she looked more like an actress than a scientist. But despite this prejudice, she built a formidable reputation as a nuclear physicist. Soon she was being compared to Marie Curie, the Polish chemist who had revealed the first secrets of radioactivity and the woman she admired more than any other.

By the mid-1950s, Wu was conducting experiments with beta decay at her low-temperature lab in Washington DC. Two of her fellow Chinese scientists, the theorists Frank Yang and Tsung-Dao Lee, suggested she look for something quite unexpected: she should ask the universe if it could tell the difference between left and right. Imagine the universe in a mirror, where we reverse all directions of space, left and right, up and down, forward and backwards. Would physics behave differently? At the time, most people didn't think so. After all, an electron would still be drawn towards a proton and repelled by other electrons. The Earth would still be bound in an elliptical orbit around the Sun. There would still be death and taxes. But when Wu performed the experiment that Yang and Lee had suggested, she noticed that beta decay always spat out left-handed electrons. When facing the direction of travel, a left-handed electron appears to spin anticlockwise, while a right-handed electron spins clockwise.[3] Wu's result proved that our universe could tell the difference between left and right, between clockwise and anticlockwise. It said that if you entered the mirror world, physics would change. It's not that everything would change; gravity, electromagnetism and the strong force would all behave as they did before. But the weak force? That would be different.

Yang and Lee were quickly awarded a Nobel Prize for the discovery, although Wu's contribution was inexplicably overlooked. The two theorists recognized the travesty of this decision and repeatedly nominated her for future prizes, without success. After Wu's ground-breaking experiment, left- and right-handed mattered, and that meant Fermi's theory needed some attention. Harvard physicist Robert Marshak and his Indian student George Sudarshan cooked up a universal recipe for the weak force known as V-A theory. It was similar in spirit to Fermi's idea, but it behaved differently to its mirror image. It also worked just as well for decays involving electrons as those involving their heavier cousins, the muons. Although there is no doubt that Marshak and Sudarshan discovered the theory first, much of the credit went to the odd couple from Caltech. Feynman and Gell-Mann had been developing similar ideas at around the same time and were actually first to publish. They were also a little noisier than their friends from Harvard. The rivalry left some bad blood. After Feynman had given a characteristically stylish talk on their work at the American Physical Society, Marshak grabbed the microphone. 'I was first!' he cried. 'I was first!' Feynman replied, deadpan, 'All I know is that I was last.'

Just as in Fermi's theory, in V-A theory, the forces act over infinitely short distances, particles kissing at a point. But we know that's not how forces really work – there is always a carrier. So how did V-A emerge as the right theory with such experimental success? Think of a Hollywood fashionista blowing a kiss to a friend, the kind where the lips don't quite touch. If the fashionista likes to get closer than most, it might look from a distance as if they did touch. V-A theory emerges in a similar way. It might look like the particles touch, but that's only because the carrier doesn't carry the force very far – because it's heavy.

So what is this heavyweight carrying the force? It turns out there are actually three particles that can carry the weak force, all of them heavyweights and all of them spin one. Two of them – the W bosons – were actually identified before V-A, by another American, Julian Schwinger. Schwinger was from the same generation as Feynman and, as giants of theoretical physics, the two were often compared. Feynman was boisterous and intuitive; Schwinger was careful and intricate. In Fermi's theory, the neutron turns into a proton by spitting out an electron and an antineutrino. Schwinger wanted to squeeze his new boson, like a gooseberry, right into the middle of that process, to stop the four particles kissing. In other words, he wanted the neutron to turn into a proton by first spitting out a negatively charged W boson, as in the figure below. Other processes involved a positively charged W boson, so he had two W bosons in total.

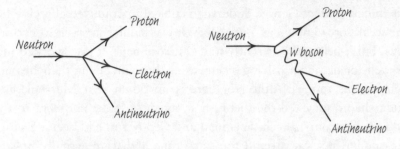

Fermi's picture   *Schwinger's picture*

*A cartoon of neutron decay. On the left you have Fermi's picture,
with the neutron instantaneously decaying into the three other particles.
On the right you have Schwinger's version, with the heavy
W boson squeezed into the middle of the process.*

Although they were taking slightly different steps, electromagnetism and the weak force appeared to be dancing in the same ballroom. In some respects, it was the dance of electric charge. On the one hand, you had the electromagnetic force moving charges through space, electrons pushing on electrons and pulling on protons. On the other hand, you had the weak force, which was able to *change* electric charge – it could shapeshift electrically neutral particles like the neutron into positively charged protons. This also meant that the weak force was carried by particles that had their own *electric* charge – that felt the electromagnetic force! Could it be that electromagnetism and the weak force were different sides of the same coin? Could the W bosons and the photon be packaged into a bundle, taking two fundamental forces of nature and combining them into a single unified force?

Schwinger certainly thought so. He tried to sew the two forces together. It was reminiscent of an artist's attempt to sew together patterns on the walls of the Alhambra, although as we saw at the beginning of the previous chapter, symmetries are special. If you want to hang on to them, you can't just sew anything together. It's why there are only seventeen patterns that ever appear on the walls and floors of ancient Islamic palaces. And it's why Schwinger couldn't quite sew together a photon and a pair of W bosons. In the end, there was too much imbalance – one boson (the photon) was electrically neutral and the other two were electrically charged. For the pattern to work – for symmetry to survive – there needed to be another neutral boson. This is what we now call the Z boson. It was a kid from the Bronx, Sheldon Glashow, who noticed that this was the missing ingredient. Glashow was a PhD student of Schwinger's, although, from the comments in his paper, it is clear he was also inspired by conversations with Gell-Mann.

Things were coming together – literally. The weak force and electromagnetism were merging into a single uber-force carried by four bosons: the photon, a pair of W bosons and the Z boson. The photon was responsible for electromagnetism, whereas the W and the Z bosons were responsible for the weak force. As with the strong force, the underlying structure was more or less the same as the one suggested by Yang and Mills a decade earlier – the one which had so upset Pauli in the seminar at Princeton. Glashow had opened the door to a unified theory of both electromagnetism and the weak force. By the end of the decade, Glashow's friend from the Bronx High School, Steven Weinberg, had applied the final touches on a theory now known as electroweak theory. At first it was ignored, but a few years later the Dutch pair of Gerard 't Hooft and his adviser, Tini Veltman, showed that it all made perfect mathematical sense and, at that point, it took off. The unification of electromagnetism and the weak force was the physics equivalent of the fall of the Berlin Wall. It was the moment that two theories became one, united into something more powerful and profound. This had happened before in physics, of course, for example when Maxwell had united electricity and magnetism, or even earlier, when Newton had connected the motion of the planets to that of a falling apple. The development of electroweak theory stands alongside those historic triumphs of Maxwell and Newton. It really was bloody marvellous.

When Weinberg moved the short distance from MIT to Harvard in 1973, he inherited an office recently vacated by Schwinger. Schwinger had left a pair of shoes, which Weinberg interpreted as a challenge: *think you can fill these?* I have no doubt that he did. That same year, the wonderfully named Gargamelle Bubble Chamber at CERN produced evidence for a weak force carried by a neutral particle, something that Weinberg's electroweak theory had predicted through the Z boson. Inevitably, Weinberg and Glashow joined Schwinger in the pantheon of Nobel Laureates.[4]

The two boys from the Bronx – Weinberg and Glashow – had been led by the guiding hand of symmetry, but there is something about electroweak theory that should worry you. I told you that the W and Z bosons were extremely heavy. They had to be because the weak force can only reach a tiny distance – around a billionth of a billionth of a metre, or about 1 per cent of the diameter of a proton. That might seem perfectly OK, but in the previous chapter we also learned that symmetry was zero and, in the case of forces, how it led to them being carried by particles of vanishing mass. So, if we live in a universe that is guided by symmetry, why does it make room for heavyweights like the Ws and Z? Why don't they have vanishing mass, as symmetry would demand?

It's time to bring in the Higgs.

The Higgs boson

A Higgs walks into a church.

> 'What are you doing here?' asks the priest.
> 'I'm here to give mass,' replies the Higgs.

I'm really sorry. I know it's a terrible joke. But what about the physics? You may have heard it said that the Higgs gives the universe its mass. Well, that just isn't true. Take the book you have in your hand, or Justin Bieber, or even a nematode worm wriggling in the soil. All of these things are heavy – they have mass – but where does that come from? Barely any of it comes from the Higgs – less than 1 per cent in fact. Thanks to Einstein's poetic equivalence between mass and energy, everything you see around you gets its mass from energy. It's the energy stored in the bonds of nuclear physics, in the chains of gluons holding together protons and neutrons. If the bathroom scales show a few kilograms heavier than you'd like, blame the gluons, blame energy, blame the doner kebabs on a Friday night. Just don't blame the Higgs.

Now everything I've just said is true about books and Bieber and nematode worms. But if we are interested in fundamental particles like the W and Z bosons or the quarks and leptons, things are a little different. The weight that they have really is down to the Higgs. We know that symmetry is zero and, when it comes to a force carrier, symmetry tells us it can't have mass. It's why the photon and the gluon are massless. For there to be heavyweights like the W and Z bosons, we must kill the symmetry.

Glashow knew this. He took the symmetry that had guided his ideas and, at the end of the calculation, he destroyed it. He smashed it to bits. But there is another way – a gentler way. To give the W and Z bosons a mass, you don't need to destroy Glashow's symmetry – you just need to *hide* it. Symmetries are hidden through a process known as *spontaneous symmetry breaking*. It's a terrible name, as the symmetry is never really broken, just hidden, but let's not dwell on semantics. Instead, I will read you a fairytale.

Once upon a time there was a princess with beautiful long, golden hair. Her name was Rapunzel and she had been locked in a tower in the middle of the woods by an evil witch. One day, a physicist was passing by and he saw Rapunzel. 'She will be perfect for my experiment,' he thought, so he took her off into outer space. When they were in the vacuum of space, far away from Earth's gravity, he noticed that her golden locks stretched out equally in all directions. This was the moment he had been waiting for. He spun her this way and that, through any angle he chose, but her hair didn't change. It always pointed in all directions. This was nature's way of telling him that the laws of physics didn't care about rotation – that they had a rotational symmetry. Not long after, he brought Rapunzel back down to Earth and repeated his experiment. The symmetry was gone. When he rotated the princess, her hair would change so that it always fell towards the ground. Of course, in time, he understood that the symmetry hadn't really gone – the underlying laws don't actually care about rotation. It was hidden from view by the Earth's gravitational field pulling down on Rapunzel's hair. In the story, the symmetry is evident in the vacuum of empty space, but in the gravitational field of the Earth it is hidden.

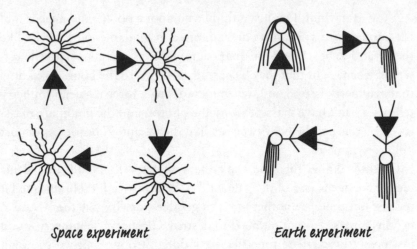

Space experiment   *Earth experiment*

Rotating Rapunzel in outer space and on Earth.

At the beginning of the 1960s, a brilliant but reserved Japanese physicist by the name of Yoichiro Nambu saw that you could play the game the other way around: sometimes, it can be the vacuum itself that hides the symmetry. Almost fifty years later, his insights would earn him a trip to Stockholm and a share of a Nobel Prize. Most people have an image of the vacuum as a desolate, empty place, where all the fields are vanishing. That is often the case, but, as Nambu realized, it doesn't have to be. By definition, the vacuum is the most relaxed of all the quantum states – the one with the lowest energy. Think of a wild house party where everyone is dancing and the house is filled with energy and excitement. Clearly that is not a very relaxed state and you certainly wouldn't call it the vacuum. Later on, when everyone has passed out, the house is in a lower-energy state. You could lower the energy even further by throwing everyone out. You could remove all the furniture. You could suck out all the air. You could empty every single quantum field. Perhaps that would be the vacuum. Maybe, but Nambu and his Italian collaborator, Giovanni Jona-Lasinio, showed that sometimes you could lower the energy some more. In their clever model of protons and neutrons, the fields weren't actually empty in the vacuum. They were filled throughout space and in such a way that certain symmetries were hidden.

Nambu and Jona-Lasinio's model may have been the prototype, but if we really want to understand how the vacuum can hide a symmetry, we should play with a simpler model – like the one involving a Higgs. We can get an intuitive picture of what is going on with a bottle of wine. First up – my favourite part – you have to empty the bottle. When you have done this, have a look at the base. You'll notice that the glass is shaped like a motte and bailey castle – a tiny hill in the middle, surrounded by a small moat. If you spin the bottle while it's sitting on the base, it doesn't really change, on account of its rotational symmetry. Now tear off a piece of the cork and drop it inside. There is a very, very small chance that it doesn't land in the moat but on top of the tiny little hill. At this point, you could gently rotate the bottle as before and, provided the cork didn't fall off, the symmetry would survive. Now, I expect that the cork actually landed somewhere in the moat. If this is the case, the symmetry is spoilt. Spinning the bottle, the cork spins too, and the picture changes. By choosing to settle in the moat, the cork appears to have broken the symmetry.

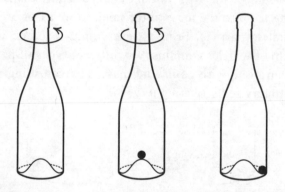

Spontaneous symmetry breaking with a bottle of wine.

The piece of cork is just like the Higgs field, and the bottle is its so-called 'potential', analogous to an electric potential or a gravitational potential – it controls what happens to the Higgs when you feed it energy or take energy away. We can read off the *size* of the Higgs field by measuring how far the cork is from the axis running straight down the middle of the bottle. In other words, if the cork is on top of the little hill, the Higgs is zero; if it's in the moat somewhere, it's non-zero. In this picture, we can also read off the energy stored in the mass of the field – it's just the height of the cork resting on the bottle. That means the lowest energy state is when the cork lies somewhere in the moat. As expected, the Higgs settles into a vacuum where its value is non-zero and the symmetry appears broken.

Except it's not *really* broken – it's just hidden.

To reveal the underlying symmetry, we need to tease out a zero. It turns out it's hiding in the spectrum of particles. Remember, a particle is just a wiggle about the vacuum – in this case, the wiggles of the cork. There are two ways in which you can wiggle the cork: you can wiggle it away from the moat or you can wiggle it along the moat. If you wiggle it away from the moat, you move it up the side of the bottle. The height of the cork tells us the amount of energy stored in the mass of the field, so this sort of wiggle is the kind we would associate with a particle of mass. In the story of the real Higgs boson, this is the heavy particle that was eventually found by smashing together protons in a tunnel at CERN. However, if you wiggle the cork *along* the moat, the height of the cork doesn't change. That means there is no energy being fed into the mass of the field, so we associate the wiggle with a massless particle. Bringing everything together, we see that the spectrum of wiggles contains two different types of particle – one with mass and one with *vanishing* mass. The vanishing mass is the hidden symmetry rediscovering its zero!

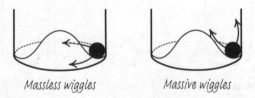

Massless wiggles   Massive wiggles

In 1962, Cambridge scholar Jeffrey Goldstone teamed up with Steven Weinberg and the Pakistani physicist Abdus Salam to prove that whenever you tried to hide the symmetries in a vacuum, the symmetry would always bite back and you would always find a massless boson. This was known as Goldstone's theorem, and it was nothing short of a disaster. The whole point of spontaneous symmetry breaking was to make a *massive* boson like the Ws or the Z, not massless ones like Goldstone's.

Particle physicists were ready to throw in the towel. Encouragement came from an unlikely source, an American condensed-matter physicist who didn't care a great deal for the microscopic dance of individual particles. But, in his own words, Phil Anderson was also 'a thoughtful curmudgeon' who had some experience with hidden symmetries from his work on superconductors. In his view, what everyone needed to remember was that the Ws and Z were *gauge* fields and the problematic symmetry was a *gauge* symmetry. As we saw in the previous chapter, that just means you can apply the symmetry at each and every point in space and time. When the symmetry is out in the open, we know that the corresponding gauge boson is guaranteed to be massless. But if it is hidden, the gauge boson should be able to acquire a mass. Aside from the weight, Anderson pointed out a crucial difference between massless and massive gauge bosons – it's in the number of working parts. A massless gauge boson has just two, like the two polarizations of a photon, while a massive gauge boson happens to have three. Anderson wondered if the extra working part came from the missing particle Goldstone had predicted. In the real world, whenever symmetries are broken, it's not that the Goldstone bosons aren't there. They are there, but somehow they get absorbed into the heavy Ws and the heavy Z. They become part of them, hiding inside, giving them just the right number of working parts.

Anderson didn't offer any details. His arguments were intuitive and set in a simple world where he didn't need to worry about Einstein and the pace of relativity. Many particle physicists believed that this was going to be the stumbling block – that when relativity was properly taken into account the whole argument would fall apart completely.

The final breakthrough came in a trinity of magnificent papers all submitted to the prestigious academic journal *Physical Review Letters*, more commonly known as *PRL*, between June and October 1964. They were written by the six wise men - Brout and Englert, Peter Higgs and Guralnik, Hagen and Kibble – five of whom would be gathered in CERN almost half a century later, waiting to hear confirmation of their work. The details were more or less as Anderson had hoped, but this time with relativity turned on. Whenever the Higgs field tumbled into its vacuum like the cork in our empty bottle of wine, the symmetry would be spoilt. The Higgs would set about giving mass to the gauge boson, and there was nothing Goldstone or his pesky bosons could do to stop it. People often say that the gauge boson 'eats' the Goldstone boson. It sounds like bosonic cannibalism, but that really is how it's allowed to get heavy. Goldstone's particle is gobbled up by the gauge boson, giving it the extra working part it needs to be massive.

The two Belgians, Robert Brout and François Englert, were first to publish, although they knew nothing of Anderson's idea. In a sense, there were two stories to tell: the story of the gauge field and the story of the field that was breaking the symmetry. Brout and Englert focused on the gauge field. Peter Higgs, a Geordie from the north-east of England, focused on the symmetry-breaker – *the Higgs*, as it is now known. He showed how his symmetry-breaker split into two parts: one that was eaten by the gauge field, giving it mass; the other a massive particle in its own right, like the cork wiggling up the side of the wine bottle. When people talk about the particle discovered at CERN – the Higgs *boson* as opposed to the Higgs *field* – this is the wiggle they mean. Initially, Higgs sent his paper to another journal, *Physics Letters*, which had published some of his previous work, but they rejected it. 'It did not warrant rapid publication,' they said. Higgs promptly took his paper to *PRL*, where it was reviewed by Nambu. It wasn't rejected a second time.

Meanwhile, Carl Hagen had gone to London to visit his old friend from MIT, Gerald Guralnik. At the time, Guralnik was working as a postdoc at Imperial College, where Tom Kibble was a young member of staff. Hagen's visit spurred them into taking on Goldstone's theorem: how to hide a symmetry and avoid the cursed Goldstone boson. Just as Guralnik and Hagen were about to post their solution to the journal, Kibble walked in, brandishing the new paper by Brout and Englert, and another by Peter Higgs. On closer inspection, they decided they hadn't been scooped. These other papers hadn't unpicked Goldstone's theorem, as they had, nor had they taken into account the quantum side of the story.

At first, no one took any real notice of any of the papers, but Kibble, especially, kept pushing. He unpicked more of the details and by 1967 he had set Weinberg up with all the right ingredients to finish off the unification of electromagnetism and the weak force. Weinberg saw that he needed to give mass to three gauge fields – two Ws and a Z – so that meant he needed a more exotic Higgs with at least four working parts. Three of those would be eaten, giving mass to the gauge fields, and a fourth would be left over. That would be the heavy Higgs boson whose discovery would be announced on 4 July 2012.

When the Nobel Prize was due to be announced the following year, many of us expected it to be given to the authors of those papers from 1964. After all, five of the six wise men were still alive – only Robert Brout had passed away, a year before the Higgs was dis- covered. No one could really choose between the five – not fairly, anyway – so there was some speculation that the Nobel Committee would relax its three-person rule. They didn't. Guralnik, Hagen and Kibble all missed out.

It was disappointing. By that time, I'd got to know Tom Kibble – Sir Tom, as he was later known. I would often see him at UK Cosmology meetings, the regular gathering of cosmologists working in the UK. Nowadays, the meetings attract up to a hundred participants, but they started out as a dozen or so people exchanging ideas in Tom's office at Imperial College, London. Tom Kibble was a giant of physics and a true gent. He never sought the limelight and always preferred to celebrate the achievements of others ahead of his own. But of the six wise men who developed our understanding of the Higgs, he was, in my opinion, the wisest of them all. More than any of the others, he built on those original ideas and, in the end, he left his mark on not one but two Nobel Prizes.[5]

THE HIGGS MECHANISM

How to give mass to a gauge boson in 8 easy steps

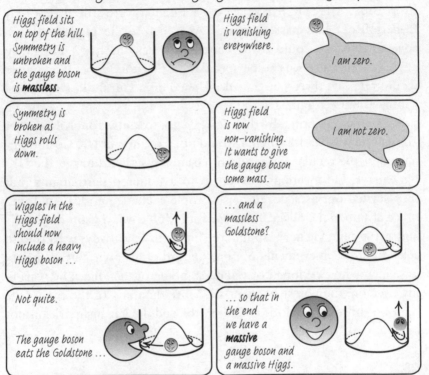

TECHNICALLY, IT'S NOT NATURAL

The Higgs tricked us. For so long, it let us believe that electromagnetism and the weak force were different. It hid from us the symmetry and beauty of electroweak theory and, as a result, the W and Z grew too heavy to penetrate our macroscopic world. It left us with the photon and an electromagnetic force that we've come to rely on. Most of our favourite devices depend on electricity and magnetism, or communicate via radio transmission. We need the electromagnetic force to scan TikTok on our mobile phones, to keep our food fresh in the fridge or to listen to our favourite songs on Spotify. Our day-to-day existence is unmistakably an electromagnetic one, not a weak one or even an electroweak one – and that is down to the Higgs.

It wasn't just the W and Z that grew fat off the Higgs and its broken beauty. So did the quarks –up and down, strange and charm, top and bottom. And the leptons: electrons, muons, taus and neutrinos. The story of how they got their mass is best told with a clever analogy that dates back to 1993 and the moment scientists from CERN were pitching to the British government for support to build the Large Hadron Collider. William Waldegrave, the Cabinet minister who was responsible for science at the time, had been struggling to understand the physics of the Higgs boson, so he challenged the team to come up with a one-page analogy that would make it more accessible. He even offered a bottle of vintage champagne for the best explanation. In the end, the British government backed CERN with financial support and a bottle of Veuve Cliquot 1985 was presented to David Miller from University College London for his brilliant analogy.

Here it is in my own words (with some creative liberties taken). There is a corner shop near my house run by a man called Dave. He's a friendly enough chap, but he's not especially well known outside our village. One day, Dave finds himself in a room with global superstar and musician Ed Sheeran. Dave doesn't really like celebrities and the atmosphere is a little tense – both men decide they want to get out. Now, as it happens, Ed and Dave have a very similar build and will naturally accelerate through the room at around the same rate. If the room is empty, they will both make it across in roughly the same time.

It's a symmetry of sorts, based on their physical similarities. But if the room is filled with hundreds of screaming Ed Sheeran fans (much to Dave's irritation), the symmetry is destroyed. Both men will be held back by the fanatical hordes but, in Ed's case, the effect is far more dramatic. He is constantly asked to sign autographs and stop for self-ies, whereas Dave is able to weave his way through without anyone paying him too much attention.

Ed and Dave are the quarks: Ed is a top quark and Dave an up quark; the hordes of fans are the Higgs field. As you might imagine, the fans will interact far more with their favourite singer than with a shop-keeper from Nottingham. When they fill the room – in other words, when the Higgs is 'switched on' – they slow Ed down far more than Dave. In a way, they make it seem as if he is heavier – they give him more 'mass'. So it is with the top quark and the up quark. The top inter-acts more strongly with the Higgs field, so when the Higgs is switched on, it gets a bigger mass. There is even a Higgs boson in this analogy. You can think of it as a ripple of excitement running through the fans. Perhaps they hear a rumour that Ed is going to sing and start whisper-ing the news to one another, gathering around in gossiping clusters. Those clusters move across the room with the rumour like a Higgs boson moving through the tunnels beneath the mountains at CERN. If you have more fans packed into the room, the clusters move more slowly, as there are more people to tell. This is like the Higgs interacting with itself, slowing itself down, giving its ripples a little more 'mass'.

Finding a Higgs like ours is like finding a snowman in the fires of hell. It *can* happen, but it really *shouldn't* happen. Suppose you took a cube of ice somewhere hot. I mean really hot, like an oven, or the fiery abyss of eternal damnation. You wouldn't expect it to carry on being an ice cube for very long. The problem is there is too much ambient thermal energy. As the air molecules bounce off the ice they will pass on that energy and the cube will melt. There is a very slim chance that this won't happen – that the molecules will miraculously keep missing the cube and the ice will survive. But it's very unlikely.

The story of the Higgs is quite similar. There is an ambient *quantum* energy that wants to make it far heavier than it is – as heavy as a fairy-fly! The quantum energy comes from virtual particles – the ones you can never hold in your hand. Remember, quantum fields are always talking to one another and, for a particle, this can sometimes cause an identity crisis.

To understand this a little better, forget the Higgs, at least for a moment. Imagine you have a photon that travelled from London to Paris. Feynman has already told us that any particle moving between two points will explore every possible path – it will go direct, it will go via the shops at the top of your road, it will even go via Andromeda. However, we also know that the photon can turn itself into an electron–positron pair, and vice versa. Can we really be sure that the photon spent the entire journey from London to Paris dressed as a photon? For a moment or two, could it not have changed into an electron and a positron before changing back? The answer is a resounding yes! Quantum mechanics creates this uncertainty – it forces us to explore every possible path, including paths where the particles undergo a costume change.

Picture a businessman making the same trip as the photon. He leaves London in a suit from Savile Row and always arrives in Paris wearing exactly the same suit. There is a chance he wore it for the whole of his journey. And there is a chance he didn't. There may have been a few moments when he wore a football kit or a cocktail dress. You never actually know. Quantum mechanics is a game of probability. If there is any chance the photon will spend some of its time dressed as an electron and a positron, you need to factor that in. You should think of

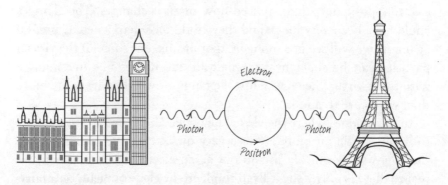

A Feynman-style cartoon of a photon travelling from London to Paris spending some of its time dressed as an electron and some as a positron.

these alternative outfits as virtual particles, never to be seen by anyone, never to be caught or intercepted, but ultimately leaving their mark. And we have felt that mark. Virtual electrons and positrons cause the energy levels in hydrogen atoms to split – something that was measured by Willis Lamb in 1947.

So, what does all this mean for a Higgs boson? Well, like the photon, if you ask how a Higgs boson went from London to Paris, you cannot assume it spent the whole time dressed as a Higgs. There is a chance it spent some time dressed as a quark or an electron or some other field we don't even know about yet. And all of these will leave their mark.

What kind of mark? Well, all those costume changes can give the Higgs a bit of a weight problem. Because there was a chance it could have spent some time masquerading as an electron and a positron, the Higgs will want to feel their weight. Intuitively, you might imagine the Higgs as being weighed down by the size of its wardrobe. The virtual electrons and positrons it changes into provide a sort of quantum medium, pulling on the Higgs as it tries to move around. With a suitcase full of these virtual particles, the Higgs gets heavy. The question is: how heavy?

If the virtual electrons and positrons weighed the same as real electrons and positrons, we'd have nothing to worry about. Real electrons and positrons are a hundred thousand times lighter than the Higgs – a suitcase this light would barely make a difference. But with *virtual* particles there is more to worry about. It boils down to the fact that we haven't said how *long* the Higgs spends masquerading as an electron–positron pair, or indeed how often it changes. The changes could have been very quick and they could have happened again and again. As we will see in a moment, that means that some of the virtual particles can be super heavy. Quantum mechanics fills the suitcase with these virtual heavyweights, weighing down the Higgs far more than we'd care to admit.

To understand where the virtual heavyweights come from, we need to think a little bit more about a very quick costume change. When the Higgs skips quickly in and out as an electron–positron pair, we realize there is only a short-lived ripple in the electron field. But short-lived ripples can mean really big energies, thanks to Heisenberg's uncertainty principle:

$$\Delta E \Delta t \geq \frac{\hbar}{2}$$

Remember my old friend Phil Moriarty chugging on his guitar in the chapter 'A Googolplex', where the shortest sounds tickled the greatest range of frequencies? It's the same with the transient electron and positron – the shorter the cameo, the greater the energies they can reach. You can now think of them as virtual particles filling the suitcase with these huge energies or, equivalently, a huge amount of mass, weighing the Higgs down more and more. If you allowed electrons and positrons to pop in and out almost instantaneously, they could touch energies that exceeded Graham's number or TREE(3) in whatever units you cared to use, and there would be no limit to how heavy the Higgs could get. But this is taking it a little too far. We can't really make sense of a Higgs changing into an electron and a positron just for an instant. That's much too quick – it would destroy the fabric of space and time. When we played the Game of Trees in chapter 'TREE(3)', we learned that you can't really do anything quicker than a Planck time, which is around 5×10^{-44} seconds. But that's still a very quick time. If we allow the Higgs to change in and out of the electron field that quickly, there will be huge uncertainties in the energy. When you sit down and work out how much mass that dumps into the suitcase – how much it feeds back to the Higgs – you find that it's very close to the mass you'd expect to find in a quantum black hole. Or a fairyfly.[6]

But the Higgs is nowhere near that heavy. In fact, its true mass is 0.00000000000000001 times smaller. Something must be badly wrong in our thinking. We know from experiment that virtual particles leave their mark on the energy levels of hydrogen, and we also expect them to leave their mark on the Higgs. So why don't we see all that extra mass? Say it quietly, but to resolve this puzzle we physicists will often resort to cheating. We just assume that there is more to the story – that there is another source of mass, something inherent in the Higgs itself. When the Higgs adds this mysterious new ingredient to the enormous mass it gets from the suitcase of virtual particles, we have to assume that it comes with the opposite sign and there is a miraculous cancellation. We said at the beginning of this chapter that it was like trying to balance a herd of African elephants against a herd of Indian elephants. Let's make that analogy a little more precise. Suppose you had a herd of around two hundred elephants whose combined weight was a million kilos. You then demand that the other herd weigh exactly the same, give or take the weight of *a single eyelash*. That's the kind of balancing act we see with the Higgs.

It's just not natural.

At this point, some of you will be shouting at me. Everything I've just said about the Higgs – about it changing outfits and gaining weight – couldn't I have said the same about the photon? Shouldn't the photon also weigh as much as a fairyfly? No, it shouldn't, and the reason is quite beautiful: it's because of its *symmetry*. We know that the photon has vanishing mass, thanks to the symmetry of electromagnetism. You might think that quantum mechanics would mess that up – that it would force all this mass on to the photon, destroying the symmetry. But here is the thing: if a symmetry is there – truly there – quantum mechanics will leave it intact. It's as if it is enchanted by its beauty. When you sit down and calculate how much mass is fed to the photon, from electrons, positrons or any other particles, you find that the answer is always zero. The symmetry and beauty are never destroyed.

The trouble with the Higgs boson is that there is no symmetry to protect its mass in the same way. It lies there at the mercy of quantum mechanics, a bubbling bonanza of virtual particles feeding it more mass than it can ever hope to stomach. To save itself, it has to perform this ludicrous balancing act, like a herd of elephants matched to within the weight of an eyelash.

THE SCARLET PIMPERNEL

It's known as the *hierarchy problem*. Why is there such an enormous difference – a hierarchy – between the mass of the Higgs measured by CERN and the enormous mass we expected it to gobble up from quantum theory? Perhaps we can take some inspiration from the electron. After all, there was a time when it too had something of a weight problem. It was before we knew much about quantum theory, when the electron was just another charged particle. Back then, the best way to calculate its mass was to figure out the energy stored in its electric field (remember, energy and mass are one and the same thing). The trouble with this is that the electron charge is usually assumed to be buried inside a single point, so when you calculate the energy stored in the electric field you actually get something infinite. That sounds ridiculous and, of course, it is. If all the electrons in your body were infinitely heavy, you wouldn't be able to move. But worse than that, you'd tear apart the fabric of space and time.

As we have already seen, we can't meddle with spacetime over infinitesimally small distances. As an alternative, perhaps we should imagine the electron charge as being stored inside a small ball whose radius is the Planck length – the smallest length you can really get away with. Well, that doesn't help much either – it makes the electron as heavy as a fairyfly, which is still way too heavy. If you insist on calculating the mass in this old-fashioned way, you need to imagine the charge smeared across a much larger ball, about a billionth of a millimetre in diameter. Then you will get the right answer – around 10^{-30} kilograms. If you want to make the ball any smaller, you need something new: a brand-new theory with new ingredients. You need a theory of quantum fields with a dash of new particle – the positron.

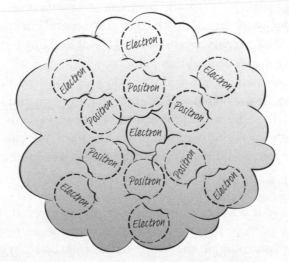

Cartoon showing a point-like electron surrounded by a cloud of virtual positron and electron pairs. These smear out the charge and make the electron seem bigger than it really is.

Once the positrons have entered the game, you can shrink the electron right down to the Planck length. A cloud of virtual positrons and electrons act to surround the electron, as if they were smearing out its charge over a much larger radius, as shown in the figure. As with the Higgs boson, the electron is fed mass by these virtual particles, but the effect is nowhere near as severe. In fact, if we imagined the electron without any mass at all, the situation would be the same as for the photon – virtual particles wouldn't be able to feed it any more mass. As ever, symmetry protects. In this case, the symmetry is a symmetry with a blemish – an approximate symmetry. That's why the electron has some mass but not too much. If you imagined a world with lighter electrons, the blemish would be smaller and the symmetry would be closer to perfection; if they were massless, the blemish would be gone completely.

So, what is this clever little symmetry? In electrodynamics, we said we had the freedom to turn an internal dial, spinning the mathematical objects we use to describe the electrons and positrons. However, this is too perfect to be the symmetry we're after. Remember, we're interested in something with a blemish – something that would only be perfect in an imaginary world of massless electrons. Well, such a symmetry does exist. It's known as *chiral symmetry*. Don't worry about the jargon. It's basically another version of the internal dial, only it turns slightly differently for particles spinning in a clockwise or an anticlockwise direction. This is a very general trick that doesn't just work for electrons. Chiral symmetries will stop any fermion being overfed the calories of quantum theory.

While that's really great, it doesn't do a lot for a particle like the Higgs boson. The problem is, it doesn't have any spin and so the symmetries are the same whether it has vanishing mass or the mass of a fairyfly. The Higgs is incapable of protecting itself, but could it have a guardian angel? Could there be something else that protects it?

Yes: the Higgsino.

Imagine a world where no one is single, where everyone is matched to a perfect partner. It seems fanciful, but this could be happening right under your nose, in the microworld of particle physics. I want you to imagine every boson hooking up with a brand-new fermion and every fermion with a brand-new boson. In other words, I want you to double the number of fields. That might seem extravagant, but underpinning all this is a new symmetry – a so-called *super*symmetry – that wants to perfect every single match. The idea is that if a boson and a fermion hook up, they should have certain things in common for their relationship to work – including their mass and their charge. This family of new particles are known as the superparticles.

How does this help the Higgs? The Higgs is a boson, so it hooks up with a new fermion known as the *Higgsino*. To make sure it's the perfect match, our super new supersymmetry *demands* that the Higgs and the Higgsino have exactly the same mass. But isn't that just wonderful? The mass of the Higgs is now tied to the mass of the Higgsino. The Higgsino is a fermion, so its mass is protected by an approximate chiral symmetry, just like the mass of the electron. It will never take on too many quantum calories. The Higgsino will never get as heavy as a fairyfly, and so neither will its partner, the Higgs. The Higgs has found its guardian angel.

We can think of supersymmetry – or 'susy', as it is lovingly known – as the most complete symmetry of space and time, as the beauty beyond all beauty. Only there is a snag – no one has ever seen such beauty.

In a world of susy, we know that the electron hooks up with a superparticle: a new boson called the selectron. The selectron and the electron are supposed to have the same mass and the same electric charge. But although we have seen plenty of electrons, no one has ever seen a selectron. It can only mean that susy isn't quite perfect. In our everyday world, it must be broken or hidden away, only to be restored when we peer at physics on the smallest of scales. Or, in other words, when we smash things together at really, really high energies. This broken symmetry forces the selectron, the Higgsino and all the other superparticles to be a lot heavier than they would otherwise be. And the more supersymmetry is broken, the heavier they get.

To find susy, we need to look for these superparticles, and that means we need enough energy to create them. Right now, deep beneath the mountains at CERN, protons are hurtling their way around the Large Hadron Collider, almost as fast as light itself. When they crash into each other, they re-create the cries of an infant universe. The energies in each head-on collision are around 10 TeV – that's what you'd get when a mosquito collides with a high-speed train. I always find that comparison a little disappointing, but remember, in the Large Hadron Collider all that energy is coming from the collision of just two unimaginably small protons. To give it the impact it really deserves, consider this: if all the protons in your body were to collide in similar fashion, they would release about twenty thousand times more energy than the eruption of Krakatoa in 1883.

When it comes to susy, what matters is that 10 TeV is around 10 million times the mass of the electron and about a hundred times the mass of the Higgs. Even so, at no point have we ever teased a selectron into existence, or indeed a Higgsino or any of the other superparticles. In the simplest models, this can mean only one thing: that superparticles are just too heavy to be produced in our collisions. This is worrying. Remember, we want to argue that the Higgsino is the guardian angel of the Higgs and that their masses are linked. But the experiments at CERN seem to suggest that the Higgsino is at least a hundred times heavier than we'd like it to be. Perhaps the Higgs need not be as heavy as a fairyfly but, in these simple models, it should be at least a hundred times heavier than it actually is. That's certainly a big improvement, but it's still a bit unnatural.

Everyone was sure that CERN would find susy. All we had to do was smash two protons together with enough gusto. It wasn't just the fact that susy saved naturalness and solved the puzzling problem of an underweight Higgs. It also solved the problem of dark matter, offering up the lightest superparticle as the perfect candidate; and it seemed to point elegantly towards further unification, towards a common origin for three of the four fundamental forces. With this hat-trick of spectacular successes, susy just had to be right. But then CERN didn't see it. People started to question the motivation for susy. They started looking elsewhere for dark matter. They started to think differently about unification.

And now there are even those who are ready to abandon naturalness.

But not all of us – not yet, anyway. If something unexpected happens, science has taught us to look for a reason. Numbers are rarely too big or too small. So, when somebody says that the Higgs boson is

0.0000000000000001 times too light, most physicists are hard-wired to seek an explanation.

We've tried plenty, and yet none has been proven right. We've tried extra dimensions. We've tried susy. We've even tried breaking the Higgs up into tiny little bits. These are all really clever ways to save naturalness, but nature doesn't seem to care. Right now, the Higgs is still the 10 million billion to one outsider that happened to win a race, and nobody really knows why.

Still, at 10 million billion to one, this is just a *small* naturalness problem. Now let me tell you about the big one.

10^{-120}

AN EMBARRASSING NUMBER

Hamburg's Haerlin restaurant was bustling with conversation. In the 1920s, this was the go-to destination for Hamburg's metropolitan elite, set within the elegant Vier Jahreszeiten, a luxury hotel on the banks of the Inner Alster. It was Otto Stern who had suggested they come here. Stern enjoyed the finer things in life – good food, good wine and good company. Wolfgang Pauli could be less choosy. Certainly, he enjoyed the glamour of the Jahreszeiten, but it was a far cry from the seedy cabaret bar in the notorious Sankt Pauli district where he had been drinking the night before. He had got into another brawl that night and was still sporting a cut above his right eye. He told Stern he had fallen over – Stern didn't need to know more. By day, Pauli lived the life of a stoic professor. By night, he was a boozing, brawling lothario.

As the two physicists finished off their brandies, Stern spoke excitedly of a new idea he'd been working on: 'I'm telling you, Wolfgang, the zero-point energy is real. I have calculated its effect on the vapour pressure of the isotopes of neon.' Pauli fixed his friend with an unmoving stare. He took a sip of brandy as Stern pressed on, 'If zero-point energy were absent, as you say it should be, the difference in vapour pressure for neon 20 and neon 22 would be enormous. Aston would have separated them with ease, but we know that he could not!'

'And what of gravity, Otto?' asked Pauli, a little deadpan. There was no reply. Pauli took out his pen and a notebook. 'Then let us calculate.' He began scribbling down some numbers as Stern sat and watched with interest. After a minute or two, Pauli looked up triumphantly.

'You see, Otto! If the zero-point energy were real, the world would not even reach to the Moon!'

The scene above is sprinkled with dramatic licence, although there are certain elements we know to be true. It is certainly true that Stern was a *bon vivant* who could only ever be found in the finest restaurants – he would sometimes fly from Hamburg to Vienna just to have lunch. This was in contrast to Pauli, who was known to frequent the bars and brothels of the Reeperbahn whenever he was out of sight of his friends and colleagues. It is also true that Stern had done his best to convince his friend of the zero-point energy but that Pauli had stood firm. Pauli's famous calculation and its withering conclusion took place at some point in the mid-1920s, as reported by two of his assistants not long after his death in 1958.[1]

But what was it that Pauli and Stern were arguing about? What is this zero-point energy?

Like Lord Voldemort, it goes by many names: it is the zero-point energy, the vacuum energy, the cosmological constant. And like Lord Voldemort, it would crush the universe into oblivion within a moment of creation. It would deny the stars and planets a chance to form. It would deny you and me the chance to be born. And yet, somehow, we made it. Nature is protecting us from this dark lord, this zero-point energy, and its thirst for Armageddon. But no one knows how. Our cosmic survival is the greatest mystery in all of modern physics.

Zero-point energy is the energy of empty space. Imagine a corner of the universe that is visited by intergalactic bailiffs. They take every- thing out – every star, every planet, every cluster of gas and every clump of dark matter. They leave nothing but a void. There are no atoms and there is no light. It is a desolate and empty place and yet in this vacuum there is something that the bailiffs cannot touch. There is energy – the zero-point energy – the energy stored in the vacuum itself. For all their efforts, the bailiffs cannot silence the vacuum. Quantum mechanics demands that it's a bubbling broth of virtual particles, con- stantly popping in and out of existence, touching the world with their energy, if only for an instant.

To understand this, I will need you to go to the kitchen and take out a large mixing bowl. Throw a little ball into it, perhaps a marble or a ping-pong ball. What do you see? No doubt the ball rolls around the bowl a little before settling down at the bottom. If you left it there and didn't touch it, you'd expect the ball to stay exactly where it is, save for a few thermal wiggles. But what if you cooled your kitchen down to absolute zero and sucked out all the surrounding air? The ball shouldn't move at all, right? It shouldn't wiggle.

It wiggles.

The reason for this is quantum mechanics and Heisenberg's famous uncertainty principle. Remember, there is always a trade-off between position and momentum. The better we know a particle's position, the worse we know its momentum, and vice versa. Let's scale down our experiment so that we throw a very light particle into a tiny bowl. If we were to say that the particle settles down, that it eventually lies motionless at the bottom of the bowl, we would have perfect knowledge of both its position and its momentum. That would violate the uncertainty principle, so something has to give. The particle must perform a little quantum wiggle. It can never settle down entirely.

With this insight, we shall go back to the empty corner of the universe. Before the bailiffs arrived it was filled with particles that conspired to make planets and stars and little green men. There were electrons and photons, quarks and gluons, gauge bosons and Higgs bosons and all the other particles we don't know about yet. These were just ripples in the fundamental fields, ripples that went away when the bailiff came and shut everything down. If you picture these fields as an ocean and the particles as the ripples on top, the job of the bailiffs is to come along and silence the ocean – to make it perfectly flat.

But the ocean is never really flat. Thanks to Heisenberg's uncertainty principle, there is always a quantum wiggle. It's the same with the fields in a vacuum – they are never completely silent. There are always tiny little disturbances. It's important to realize that these disturbances aren't real particles, because the bailiff will grab those and take them away. So, they must be virtual. In fact, they're very much like the transient electrons and positrons we saw in the previous chapter, as the Higgs made its journey from London to Paris. To recap, the Higgs left London as a Higgs, and arrived in Paris as a Higgs, but what it did in between was anyone's guess. One possibility was that it remained a Higgs for the whole journey; another was that it spent some time dressed as an electron–positron pair. Feynman told us that a particle will explore every path and every possibility. Each of these paths will leave their mark on the Higgs, giving it some mass.

It's the same with the vacuum. If we go back to the empty corner of the universe, we might see that it's empty first thing in the morning, and then again some time later. The time interval doesn't really matter. What matters is that you start out empty and finish empty, but what happens in between is anyone's guess. The vacuum could easily have changed costume, just as the Higgs did, allowing virtual particles to come in and out of existence, like popping candy. These virtual particles leave their mark on the vacuum just as they did on the Higgs. They give it mass. They give it energy – a *lot* of energy.

To figure out how much energy is hiding in the vacuum we need to break it up into tiny little pieces, like a magnificent cosmic jigsaw in three-dimensional space. As we will see, the size of the pieces will radically affect the result. If we were interested only in physics that could be seen with the naked eye, we might set the pieces to be a box just under a millimetre across. But we should be more ambitious than that. When Pauli was thinking about this over lunch, he set his jigsaw pieces to be the size of the classical electron radius, a few femtometres across. This is a much smaller distance than you could ever hope to see with the naked eye, roughly ten thousand times smaller than an atom. In Pauli's day, it was right on the edge of physics, at the boundary of what they were trying to understand.

As ever in a relativistic world, with a shortest distance comes a shortest time. If our jigsaw pieces are a few femtometres across, as Pauli imagined, the shortest time we could realistically consider is around a hundredth of a trillionth of a nanosecond. This is the inconceivable time it takes light to travel across one of our boxes. We use it to put a limit on how quickly virtual particles are popping in and out of the vacuum – we don't consider particles that pop in and out more quickly, as that would correspond to a set-up with smaller jigsaw pieces. These transient tremors feed the vacuum through an ambient quantum energy, just as they did for the Higgs. Those popping in and out the fastest will feed the vacuum with the most energy, their frantic high-frequency cameo dumping as much energy as the uncertainty principle will allow. This works out at around five trillionths of a joule[2] for every one of our little boxes. That might not seem like much, but remember, the boxes are tiny so the density is dangerously high. In every coffee cup of empty space, you would then find almost 100 thousand trillion trillion joules, enough energy to boil away all the oceans of the Earth.

But we shouldn't stop here.

It is almost a century since Pauli performed his quirky calculation and, since then, we have learned to look a lot deeper. At CERN, particle collisions have pushed the boundaries ten thousand times further than Pauli had imagined. The edge of experimental physics now lies at an unfathomably small distance of around 10^{-19} metres. If we make our jigsaw pieces this small, we can consider virtual particles popping in and out of the vacuum every billionth of a billionth of a nanosecond. The vacuum continues to devour all this quantum energy in vast quantities. An empty coffee cup would now have enough energy to blow up an entire planet, Star Wars-style, smashing it to bits, firing off planetary shrapnel at high speed towards every corner of the universe. And it could do this more than 100 billion times, wiping out every planet in the galaxy.

But we shouldn't stop there.

The collisions at CERN represent only the edge of *experimental* physics, constrained as it is by limits on funding and technology. But physics itself doesn't stop there. It goes on. It takes us right to the brink, to the point where any notion of space and time begins to disintegrate. The pieces of the jigsaw really ought to be as small as the Planck length, more than a million billion times smaller than the edge of our experiments. The implications for the vacuum are terrifying. Particles are popping in and out of empty space in every Planck time or, in other words, every 10^{-35} seconds. The ambient quantum energies become truly monstrous and the vacuum consumes them with ravenous glee. In every litre of empty space, we should find a googol *giga*joules of energy. Wow! In every coffee cup of vacuum, there should be enough energy to destroy every planet in the observable universe again and again and again and again, wiping everything out more than a trillion trillion trillion trillion times.

Do you feel frightened, knowing that these gargantuan energies could be everywhere around you, and even inside you, in the empty space between your atoms? How did you ever get this far with this monster within? The truth is, without gravity, there is nothing to worry about. It doesn't matter how much energy there is lurking in the vacuum, we cannot turn it into a weapon, exerting forces formidable enough to destroy planets. In fact, we cannot use the vacuum energy at all. This is because it's the same everywhere. To get anything exciting to happen, you need energy *differences* – you need gradients – and with the true underlying vacuum energy, there just aren't any. The energy of empty space is the zero-point, the baseline, beyond which everything else is measured. It can never be used to give an extra push or an extra pull. It simply cannot touch you, not without gravity.

But with gravity it goes wild.

With this much energy in empty space, a universe that obeyed Einstein's laws would be crushed under its own weight. Not only would it 'not reach to the Moon', as Pauli declared, it would not even reach to an atom. It would be a broken spacetime, wrapped up and twisted, extending little more than a Planck length in any direction.

Einstein taught us that it is really energy that gravitates, not mass. A photon from a distant star skirting around the Sun is bent inwards. The Sun isn't pulling on its mass because the photon doesn't have any. It pulls on its energy. In Einstein's world, all forms of energy are waltzing the gravitational waltz. Everything must dance: the Sun, the planets, you, me, an alien badger, a black-hole behemoth, and even the vacuum itself.

The energy of the vacuum is everywhere, unchanging in space and unchanging in time. This is why it's sometimes called the *cosmological constant*. Like any energy, it will curve the spacetime in which it lives. When the energy is positive, a horizon will form around each and everyone of us – a de Sitter horizon, as we saw in the chapter 'Graham's Number', representing the edge of what we can ever hope to see. The more energy hidden in the vacuum, the closer the horizon and the smaller our world. If we estimated the vacuum energy using Pauli's jigsaw, the horizon would lie at around 237 kilometres. Never mind the Moon, the universe would not even reach as far as the International Space Station. As we improve our estimate – as we make our jigsaw pieces smaller and smaller – the horizon closes in. For the jigsaw pieces as small as a Planck length, the horizon is right upon us, barely a Planck length away. This is a universe defeated by the void, crushed and crumpled by the weight of nothing.

This is not our universe.

Look around you. The horizon of our universe is not upon you. As we saw in chapter 'Graham's Number', it is unimaginably far away, at almost a trillion trillion kilometres. Ours is a universe slowly accelerating, distant galaxies being pushed apart by something unseen. We call it dark energy, but that is just a name. Most of us think it's the pressure of the vacuum – of the zero-point energy hidden away in empty space. But its push is extremely gentle. To match the rate at which distant galaxies are accelerating away from us, the energy of the void must be very thinly spread, with less than a trillionth of a joule in every litre of space. It bears no resemblance to the jigsaw estimates we made with quantum theory. A coffee cup filled with the true vacuum doesn't have enough energy to destroy a planet or boil the oceans. In fact, you'd need the energy of at least ten thousand cups just to crush a fairyfly and, as you already know, that is the smallest insect in the entire world.

This is embarrassing.

Quantum field theory – the microscopic description of particles and fields – is often billed as the most accurate theory in the history of mankind, and with good reason. Some of its predictions, like the so-called anomalous magnetic moment of the electron, have been tested and verified to an accuracy of one part in every trillion. And yet here we are, trying to predict the energy density of the vacuum with this champion theory, only to find that the true value is 10^{-120} times smaller. That's really quite a small number. If you wrote it out like a decimal, it would look something like this:

0.0000000000000000000000000000000000000000000000000000000000000 0000000000000000000000000000000000000000000000000000000000000 0000000000001

As we have seen, nature doesn't do small numbers – not without a very good reason – so why is it there? Our best and most magnificent theory predicts a vacuum filled with a googol gigajoules of energy in every litre of space, and yet nature tells us there is barely a picojoule. It's the most inaccurate prediction in all of physics. Of course, we should be grateful. If our predictions had been right, the universe would have been bent and broken by gravity, a cosmic runt barely extending in either space or time, unable to support the stars and planets needed to harbour intelligent life. But our predictions were not right. We are lucky enough to live in a vast and aged universe, where the vacuum energy is 10^{-120} times smaller than expected, where there is a tiny number we just don't understand.

This is the most embarrassing number in fundamental physics, a spectacular discrepancy between our state-of-the-art calculations and the reality of what we see around us. Einstein's general theory of relativity and quantum field theory are the best and most well-tested theories of the twentieth century, and yet when we put them together we encounter this calamity – a calamity known as the cosmological constant problem.

ALBERT EINSTEIN'S MOST DIFFICULT RELATIONSHIP

The story of the cosmological constant begins with Planck and the *Nullpunktsenergie*. The name conjures up an image of a German rock band banging out the tunes in an underground cellar in the mid-1980s. But it has nothing to do with sweat and mullets and electric guitars. It's the zero-point energy, first introduced by Planck in his *second* stab at quantum theory, in the years leading up to the First World War. We came across his first attempt in the chapter 'A Googol-plex'. This was where Planck saved us from the ultraviolet catastrophe by splitting energy up into chunks. The idea had worked out wonderfully – and it was right – but Planck didn't care for it. He was never at ease with the chunkiness and said at the time that if he could abandon it, he would. In the end, he managed to abandon half of it. In his second attempt at quantum theory, he declared that radiation still had to be emitted in chunks, but it didn't have to be absorbed that way. The lack of symmetry seems ugly to us now, but in the early days of quantum theory it felt a little less radical, a little more conservative. But there was a price to pay for this. To get his alternative quantum theory to work, Planck needed some left-over energy, even at the zero point, when you cool things down to absolute zero. He needed the *Nullpunktsenergie*.

Planck's second quantum theory would never eclipse his first for the simple reason that it just wasn't right. Nevertheless, the idea of a zero-point energy had caught the wandering eye of Einstein and his accomplice, Otto Stern. At around the same time, the German chemist Arnold Eucken had obtained some data for the specific heat of hydrogen molecules. The details don't matter – what matters is that Einstein and Stern showed that the zero-point energy could help them understand the data. But Einstein's affection didn't last. Within a couple of years, he had grown violently opposed to the whole idea of the zero-point energy. 'No theoretician,' he scoffed, 'can ... utter the words "zero-point energy" without breaking into a half-embarrassed, half-ironic smile.' It was the troubled Austrian physicist Paul Ehrenfest who had turned him.

Ehrenfest had been able to fit Eucken's data without any zero-point energy at all, using only Planck's original quantum theory – the one we now know to be right. Einstein took the view that if you didn't need something, then why bother with it, and he respected Ehrenfest. He was also a very close friend. It's worth pausing for a moment to take in Ehrenfest's story, because it is probably the most tragic in all of physics. He had been a student of Boltzmann, in the later years, when the great man was racked with self-doubt. Boltzmann's suicide came just two years after Ehrenfest had completed his doctorate. He was beginning to build his own reputation, not just as a great physicist but as the greatest teacher of his generation. 'He lectures like a master,' roared Sommerfeld, perhaps the most influential physicist in Germany, 'I have hardly ever heard a man speak with such fascination and brilliance.' But for all his brilliance, Ehrenfest was tormented by demons more devastating than those that had broken his mentor. And Einstein knew it. In August 1932, he wrote to the University of Leiden, where Ehrenfest worked. He was worried about his friend. Ehrenfest's marriage had failed and he had given up on physics. Einstein saw that he was being overcome by the darkness of depression. A year later, Ehrenfest would be dead. On 25 September 1933 he travelled to the Institute for Afflicted Children in Amsterdam to meet with his fifteen-year-old son, Wassik. Wassik had Down's Syndrome and had recently been taken out of Germany for his own protection when the Nazis seized power. When Ehrenfest met with him in the waiting room, he took out a pistol and shot him in the head. Moments later, he turned the gun on himself.

It was Ehrenfest who had turned Einstein violently away from the zero-point energy. It may well have been Ehrenfest who turned him back. Something happened in between the war years and the early 1920s so that Einstein was once again seduced by the idea. We don't really know what it was. What we do know is that he and Ehrenfest were writing to one another, Einstein suggesting that zero-point energy could account for a very curious property of helium. Whenever an element is cooled, the molecules lose their kinetic energy and the liquid phase gives way to solid. But with helium, this never happens, at least not at atmospheric pressure. It doesn't matter if you cool it all the way down to absolute zero, it will never become a solid. And Einstein was sort of right – the reason for this is related to the zero-point energy. It equips helium with an internal pressure, forcing it to expand to lower density, preventing the formation of rigid structures.

In the early 1920s, molecular chemists like Harvard's Robert Mulliken were seeing more and more evidence in favour of a zero-point energy but, with Planck's second quantum theory discredited, its origin was not well understood. That would change in 1925 when quantum mechanics finally came of age. The blossoming of quantum mechanics is a tale of two retreats. I've already told you about Schrödinger slithering off to the Alps with his mistress, conjuring up an equation that would shake the world of physics. But six months earlier, Werner Heisenberg had also fled the city, to the North Sea island of Helgoland. Unlike Schrödinger, he wasn't running from his wife – he was running from 'the blossoms and the meadows'.

Heisenberg's story lacks the tabloid sense of scandal, but it is no less important. In the late spring of 1925, he had become very ill with hay-fever and had gone to the islands to escape his allergies. The swelling around his face was so bad that when he checked into the guesthouse overlooking the dunes, the landlady assumed he had been in a fight and promised to nurse him back to health. Beyond the occasional walk or swim in the sea, there wasn't much to distract the young physicist in his island retreat. He had the freedom to think more deeply about the hydrogen atom, trying to understand the origin of its spectral lines – the chunks of energy it can absorb and emit.

His obsession with the problem soon turned into insomnia, but a breakthrough finally came to him in the early hours, on a hot summer's night. 'It was about three o'clock at night when the final result of the calculation lay before me,' recalled Heisenberg. 'At first, I was deeply shaken. I was so excited that I could not think of sleep. So, I left the house and awaited the sunrise on the top of a rock.'

Heisenberg realized that electrons in an atom didn't have sharp orbits, as Bohr had originally suggested. That seemed to be true only when the electrons were higher up, far away from the nucleus. Closer in, things were a lot fuzzier. You couldn't reliably say if the electron was in this orbit or that. Whereas Schrödinger captured this fuzziness with an intuitive picture of waves, Heisenberg used the more abstract mathematical language of matrices. But these were just two different descriptions of exactly the same thing – the sorcerer's world of quantum mechanics where everything is a game of chance.

Heisenberg's work was a tour de force. Just as Newton had invented calculus to describe the mechanics of the macroworld we see every day, so Heisenberg had invented a new mathematics to describe the microworld we cannot see. It wasn't as easy to work with as Schrödinger's theory, but it was first, and was able to capture the abstract beauty of the quantum world with fewer ingredients.[3]

In 1933, the year Heisenberg was rewarded with the Nobel Prize, the Nazis took control of Germany. They began to develop policies that targeted non-Aryans or people considered politically unreliable in the civil service. Many academics fell victim to this campaign or resigned in protest. But Heisenberg chose to be quiet in his opposition. He took the view that Hitler wouldn't be around for long and so he should just keep his head down. But then he too became a target. The Nazis saw too much Jewish influence in the abstract and mathematical approach to science that was developing in the early part of the twentieth century. When Heisenberg was lined up for a prominent professorship in Munich, he fell into the cross hairs of Johannes Stark, himself a Nobel Prize-winning physicist, as well as a rabid Nazi. Stark signed an SS article declaring Heisenberg to be a 'White Jew' and the 'Ossietzky of Physics'. (Ossietzky was a German journalist and pacifist who had been interned in a Nazi concentration camp.) It was Heisenberg's mother who intervened. Her family had connections to

Heinrich Himmler, who declared a compromise: Heisenberg would be spared further personal attack, but he would not be going to Munich.

Heisenberg stayed in Leipzig. He was not short of offers to work elsewhere, especially in the United States, but he felt a strong personal duty to stay in his home country, regardless of its politics. During the war, he took a leading role in the German nuclear research programme. Some believe that Heisenberg deliberately undermined the more sinister aspects of the programme, although this is not entirely clear. On a visit to Denmark in 1941, he had upset Niels Bohr by bringing up the subject of nuclear weapons research. Heisenberg would later claim that Bohr had misunderstood his intentions. A year later, Heisenberg met with Albert Speer, the Nazi minister for armaments, and advised against pursuing any further research into nuclear weapons. He did, however, continue to experiment with nuclear power and there is no doubt he was looking to advance Germany's scientific reputation.

While I was writing this chapter, I took a break with my family in Germany, on a farm in the Black Forest. A change to our travel plans meant we were short one night's accommodation, so I booked a room in an ancient castle on the edge of the forest, overlooking the picturesque town of Haigerloch. As luck would have it, the hotel had played a role in the history of quantum physics. In the caves beneath the castle, far away from the bombs that were raining down on Berlin, Heisenberg and his colleagues had built a nuclear reactor. It was a final desperate attempt to win the race for atomic power as the Allies closed in towards the end of the war. The cave is now a museum featuring a full-size model of Heisenberg's experiment – cubes of uranium suspended on chains in a vat of heavy water. Neutrons, slowed by the heavy hydrogen atoms, were used to split some of the uranium nuclei, firing off more neutrons, splitting more nuclei. The goal was to trigger a self-sustaining chain reaction, releasing huge amounts of atomic energy. Heisenberg and his team had been close to success – just 50 per cent more uranium in the core would have been enough for the reactor to work. By the time the Allied troops discovered the cave, Heisenberg had fled Haigerloch by bicycle, under cover of darkness. The uranium cubes were found buried in a field next to the castle.

The Allies soon caught up with Heisenberg at his Alpine home in Bavaria, which was still under German control, and took him to the

Farm Hall Estate in England for interrogation. British Intelligence secretly recorded the conversations that took place between the scientists held at Farm Hall and the transcripts were released in 1992. Although Heisenberg's reactor had been close to working, he was overheard telling the other scientists he had never seriously contemplated a bomb. 'I was absolutely convinced of the possibility of our making a uranium engine,' he said, 'but I never thought that we would make a bomb and at the bottom of my heart I was really glad that it was to be an engine and not a bomb. I must admit that.'

It was Heisenberg who first understood the origin of the zero-point energy, which emerged from his brilliant formulation of quantum mechanics. He showed that a quantum oscillator – a little quantum wiggle – could never be energy-free. The physics of fundamental particles is really the physics of these tiny wiggles. Whenever you have real particles, the wiggles are in an excited state. When you're in a vacuum, the wiggles settle down as much as the uncertainty principle will allow and, as Heisenberg showed, the energy doesn't vanish.

But is this vacuum energy *physically* real?

A gecko would say that it is, as it scampers across the ceiling. Its magical ability to walk on walls is thought to rely on *changes* in vacuum energy and the force of a quantum vacuum. It turns out that the energy of a vacuum depends on the shape of its surroundings. We know that zero-point energy comes from the ripples of virtual particles bouncing in and out of existence. Crucially, however, these ripples depend on the size and shape of the vacuum edge. You see a similar effect with ripples on a body of water – they depend on the shape of the pool, or lake, or even the ocean. If you change the edge of a vacuum, you change the virtual ripples, and that can change the zero-point energy. This means the vacuum will push and pull on the walls that surround it, to try and change the ripples and lower the energy. It leads to the so-called Casimir force, named after the Dutch physicist Hendrik Casimir, a student of Ehrenfest's. When the walls of a vacuum are far apart, the force is tiny, but if they are microscopically close, the force can be measured. (Steve K. Lamoreaux and his team at Los Alamos National Laboratory did precisely that in 1997.) In a similar vein, changes in the zero-point energy can also lead to so-called van der Waals forces between atoms and molecules. This brings us back to the gecko. Some biologists think that geckos use van der Waals forces to stick to the ceiling, thanks to the zero-point energy changing in the vacuum between microscopic projections on the soles of their feet.

These measurable effects give us confidence that the theory of zero-point energies is right, but the truth is they are measuring only parochial changes – fluctuations in zero-point energy that occur whenever we surround a bit of empty space with a wall of atoms and molecules on the foot of a gecko. Experiments like Lamoreaux's in Los Alamos tell us very little about the monster underneath – the vast reservoir of vacuum energy that underpins the entire universe. This is the zero-point energy you would still expect to find when you remove all the walls and empty the universe completely. As we have seen, this monster should be huge. It should crush the universe out of existence.

The cosmological tale of the zero-point energy began independently of its development in quantum mechanics. To pick up this particular story, we must return to the first few months of 1917, eight years before Heisenberg had revealed its quantum origin. At the time, Albert Einstein was still violently opposed to the zero-point energy and wasn't inclined to think too much about it. But he was thinking about gravity and the impact his wonderful new theory would have on the universe as a whole.

He began with a puzzle – the problem of infinite space. Could such a thing ever really make sense? To avoid this issue, Einstein preferred to think of the universe as an enormous sphere, like the surface of a ball, huge but ultimately finite. In General Relativity, Einstein's equations relate the shape and size of the universe to the matter it contains. On the grandest scales, he saw that his spherical universe was forever being pushed and pulled by the matter within. It would never settle. Einstein didn't like that at all. The idea of a universe evolving in time was abhorrent to him. His intuition demanded an unchanging world, without beginning or end, but the equations refused to play ball. He needed a fix.

Einstein saw that he could check the troubling evolution with a new ingredient – a cosmological constant – permeating all of space and all of time. He plucked this cosmological constant out of his imagination – he had no idea it could be connected to the zero-point energy of the universe. But now that he had imagined it, he set things up just so, the cosmological constant carefully balancing matter and the curvature of space so that the universe stayed still. It was an uneasy truce forged between cosmic giants on the battlefield of space-time. It was a truce that wouldn't hold.

The first warning sign for Einstein came later that same year, in 1917, under the withering attack of the Dutch astronomer Willem de Sitter. De Sitter questioned many of Einstein's underlying assumptions and showed that there were viable alternatives to Einstein's universe, both experimentally and mathematically. He imagined a universe that was so dilute, it could be treated as if it were free of all matter, leaving only the ubiquitous cosmological constant. This gave him an alternative cosmic solution – a universe shaped entirely by Einstein's cosmological term. Einstein didn't believe it could possibly describe our universe precisely because there was no role for ordinary material objects like stars and planets. To make matters worse (at least as far as Einstein was concerned), if you did throw in some stars and planets, the astronomer Arthur Eddington showed that they would fly apart, accelerating away from one another as the space in between them expanded. De Sitter and Einstein respected each other enormously and, although they debated hard, there is no evidence that Einstein ever accepted the reality of de Sitter's solution. The Einstein world and the de Sitter world became the leading cosmological models of the day.

Alexander Friedmann wasn't interested in taking sides. In 1922, this young Russian physicist decided to take the possibility of an evolving universe much more seriously and found a whole new family of solutions. In Friedman's world, there was no cosmological constant. Instead, the expansion was driven by matter, with the expansion *slowing down* as matter was diluted. Contrast this with the previous two proposals. In Einstein's world the universe was still; in de Sitter's world, there was also expansion, but it was driven entirely by a cosmological constant, forcing the expansion to speed up, to *accelerate*. Save for a few bursts of acceleration at very early and very late times, it turns out that Friedman's expanding but decelerating cosmology is the best model of our universe for most of its history.

Einstein initially dismissed Friedmann's paper as being mathematically flawed. When it became clear it was mathematically sound, he began to see its importance, triggering a change in his relationship with the cosmological constant he had introduced five years earlier. In a postcard he sent to Hermann Weyl in 1923, he declared, 'If there is no quasi-static world, then away with the cosmological term.' In other words, if you accept the idea of an expanding universe, he saw no point in discolouring General Relativity with his fix from 1917 – there was no point in having a cosmological constant. This would be the dominant view for the next seventy years as all the evidence pointed towards a universe that was expanding but slowing down, just as Friedmann had suggested. As we will see, the cosmological constant did not return until the 1990s, when astronomers began to detect hints of acceleration in the very latest stages of our cosmic history.

Friedmann would never see his model triumph. In the summer of 1925 he ate a pear at a railway station as he travelled home from his honeymoon in Crimea. The pear had not been properly washed and was allegedly riddled with bacteria. After becoming unwell upon his return to Leningrad, Friedmann was diagnosed as having typhoid and within two weeks he was dead.

It was around this time that the Abbé Georges Lemaître began to develop his own ideas. Growing up in a well-to-do Catholic home in Charleroi, Belgium, Lemaître had decided to become a priest when he was just nine years old. In the same month, he also decided to become a scientist. 'I was interested in truth,' he told *The New York Times*, 'from the standpoint of salvation, you see, as well as in truth from the standpoint of scientific certainty.' He never saw any conflict between these two aspects of his life.

Lemaître had not been following Friedmann's work, but he had read the publications of Vesto Slipher, an American astronomer who had been observing dim spirals of light known as spiral nebulae. Slipher had noticed that these spirals were moving away from us, which Lemaître correctly attributed to cosmic expansion. Rough estimates suggested that the spirals were extremely distant, so much so that some astronomers speculated that they were actually island universes, made up of millions, perhaps even billions, of stars. And they were right. Edwin Hubble was able to peer in a little more closely and identify some of the individual stars. Slipher's spiral nebulae are what we now call galaxies.

Lemaître set about solving the equations for an expanding universe, but Einstein was unimpressed. Lemaître had thrown everything into his model: planets, stars and even the cosmological constant. To Einstein, that was overkill. In an expanding world, he saw no value in the cosmological constant. As far as he was concerned, its sole purpose had been to halt the expansion and render the universe static. When Lemaître sought him out to discuss the paper at a conference in Solvay in 1927, Einstein wasn't in the mood for clemency. 'Your calculations are correct,' he praised, drawing Lemaître in, 'but your physical insight is abominable.'

Eddington was more complimentary. He saw that Lemaître's work represented the end for Einstein's static model of the universe. Although Lemaître had never said so explicitly, his calculations implied that Einstein's world was unstable. It relied too heavily on the uneasy truce between matter and the cosmological constant. If the truce were broken, albeit gently by tweaking the density of matter by a tiny amount, the universe would quickly change into something else. And one thing is certain, it would never be static.

By the end of the 1920s, Hubble had been able to accurately measure the distance to Slipher's galaxies. When this was compared to the speed at which they were moving away, it confirmed the expanding model of the universe, in accordance with the cosmologies developed by Friedmann and Lemaître, and in contrast to Einstein's original model of 1917. At this point, Einstein became more vocal in his rejection of the cosmological constant. The universe wasn't static, so it simply wasn't needed.

It is often reported that Einstein described the cosmological constant as 'the biggest blunder of [his] life', though there is some debate as to whether he actually ever said it. It is certainly true that Einstein never returned to the cosmological constant. In a review article he wrote towards the end of the Second World War, he confessed, 'If Hubble's expansion had been discovered at the time of the creation of the General Theory of Relativity, the cosmologic [sic] member would never have been introduced.' A couple of years later, he wrote to Lemaître, lamenting the ugliness of the cosmological constant and declaring that he had 'always had a bad conscience' for introducing the term. As for the 'biggest blunder', that quote was first revealed by the Ukrainian physicist George Gamow. Although the renowned American physicist John Wheeler claimed he overheard the remark in a conversation between Gamow and Einstein at Princeton, some doubts have been raised, largely on account of Gamow's character. A brilliant physicist, Gamow was also a drinker with a mischievous sense of humour. On one notable occasion, he secretly added the name of his friend, Hans Bethe, to a seminal paper written with his student Ralph Alpher on the synthesis of light elements like hydrogen and helium. The inclusion of Bethe's name meant the authors could be listed as Alpher–Bethe–Gamow, as if reading off the first three letters of the Greek alphabet. In any case, it doesn't really matter whether or not Einstein really described the cosmological constant as his 'biggest blunder'. It certainly paled in comparison to his greatest regret: the letter he'd signed to President Roosevelt in 1939, warning him that Germany might build an atomic bomb and encouraging the United States to develop its own nuclear weapons.

Lemaître did not allow himself to be disheartened by Einstein's criticism and continued to ponder the implications of the cosmological constant and an expanding universe. In a letter to the journal *Nature* in 1931 (published immediately after a discussion of the insects found in the gut of a cobra), he asked what would happen if we went back in time and imagined the universe as it was long, long ago. He realized that the energy of everything – every planet, every star, every pulse of radiation – would be packed into the tiniest space, perhaps into a single unknown 'quantum'. Lemaître was trying to resolve what we now call the initial singularity, the primeval crush of infinite density that marks the beginning of space and time. As for the cosmological constant, Lemaître never gave up on it, in contrast to Einstein. He was the first to identify it as the energy of the vacuum, but he never made the connection to the zero-point energy and quantum mechanics. Perhaps if he had, he'd have drawn Einstein back into the fold.

For the next three decades, the cosmological constant was largely ignored, even among the handful of physicists who were studying cosmology. The finest minds in the field were more interested in particles, in wrestling with the microworld and picking apart the structure of the fundamental fields. The cosmological constant had originally been championed by a priest. And it would be resurrected by a man who led the Soviet nuclear weapons programme in the aftermath of the Second World War. Yakov Zel'dovich was one of only sixteen people to have been named a Hero of Socialist Labour, the Soviet Union's highest degree of distinction, on three separate occasions. In the late 1960s, he joined up the dots in the cosmological vacuum, connecting Heisenberg's zero-point energy with the cosmological constant. It was Pauli's café calculation, dressed up with modern ideas. And just as Pauli did before, Zel'dovich identified a problem. A bloody big problem.

Zel'dovich realized that if quantum field theory were right, the vacuum should be filled with a broth of virtual particles, forever popping in and out of existence. This broth should add a sort of weight to the vacuum, filling it with so much energy and pressure that the universe would be bent into oblivion. The cosmological constant could no longer be ignored.

Half a century on from Zel'dovich's declaration, the cosmological constant problem is still here and, if anything, it's grown worse. Zel'dovich believed the true cosmological constant to be vanishing. He didn't know how it should vanish – he didn't know what could tame the soup of virtual particles – but there had to be something. A symmetry, perhaps? Thirty years later, in the late 1990s, astronomers began to see evidence for cosmic acceleration, distant supernova moving away from us at ever-increasing speeds. This acceleration looks as if it's being pushed by a cosmological constant, but it's not the cosmological constant predicted by quantum theory and the frenzy of virtual particles popping in and out of the vacuum. It's a cosmological constant that is 10^{-120} times smaller.

Although the true value of the cosmological constant raises some very difficult questions, its existence is usually billed as an unexpected triumph for Einstein. He may have ultimately discarded it, but make no mistake, the cosmological constant was his invention. The accelerating universe is also a triumph for de Sitter. As our universe expands, becoming more and more dilute, it seems to be approaching the de Sitter world, an empty and eternal universe driven on by the ubiquitous cosmological constant. But one question still remains.

Why is it so *embarrassingly* small?

THE GOLDEN TICKET

The situation is getting rather desperate. It has been almost a century since Pauli sat with Stern in a café in Hamburg and declared that the universe 'would not even reach to the Moon'. In all this time, no one has conjured up a solution to the cosmological constant problem that satisfies everyone, or perhaps even *anyone*. We know that little numbers shouldn't happen by accident and yet there it is, a cosmological constant 0.000000000000000000000000000000000000000000000000000000 000000000000000000000000000000000000000000000000000000000000 000000000000000000000001 times its expected value. Naturalness has been gloriously successful in almost every other area of fundamental physics but, in the cosmological vacuum, it is drowning.

Bohr was one of the first to try and save it. In 1948, in his opening address at the Solvay conference in Brussels, he mused over zero-point energy. Like Pauli, he knew that if gravity were to see it, it would go wild, bending space into oblivion, so, in his mind, something must be making it vanish. He imagined a perfect balance in the bubbling broth – some particles endowing the vacuum with a positive energy, others with negative, cancelling one another out. It's like being surrounded by an equal number of angels and demons. The angels will bring you gifts of happiness and joy, while the demons will take them away. If the balance is right, you are neither happy nor sad. So it might be with the cosmological constant: some virtual particles try to push it up, and others try to push it down. In the end it settles at zero.

Bohr speculated that virtual protons and electrons might compete in this way. As it happens, they don't, because they are both fermions. Virtual fermions in the vacuum soup always try to push the vacuum energy down, pushing us towards *negative* energy. But virtual bosons do the opposite – they try to push the energy up. Pauli was the first one to spot this. If the bosons behaved as angels, and the fermions as demons, then, in perfect balance, they might cancel one another out, taming the cosmological vacuum just as Bohr had imagined.

It's a nice idea. But so are magical unicorns – it's just that they have no place in our particular world. To get just the right balance between bosons and fermions, you need a symmetry we encountered in the previous chapter – you need 'susy'. Susy was the supersymmetry we imagined protecting the mass of the Higgs. The idea is that you double the number of particles, so that every boson is married to a new fermion and every fermion to a new boson. For each of the marriages to work, both particles have to have the same mass and electric charge. When it comes to cancelling the cosmological constant, this is exactly what you need. In a perfectly supersymmetric world, each virtual boson would try to weigh the universe down with vacuum energy, only for its fermionic partner to cancel out the effect. But our world is not perfectly supersymmetric. In fact, we haven't seen any sign of susy – not yet, anyway. If we break the vacuum up so that the jigsaw pieces take us to the edge of experimental physics – to the brink of the collider experiments at CERN – there is never any susy and so no chance of a miraculous cancellation of vacuum energy.

This is just one failed attempt, but the truth is there have been many. Like a siren, the cosmological constant problem entices its prey. Physicists are drawn in, determined to conquer it, to protect naturalness. But they never seem to succeed. For half a century or more, the cosmological constant problem has defied us, and those failures are weakening our resolve. There are those who believe that naturalness is already dead. In desperation, they've abandoned the old ways and sought refuge in a new way of thinking.

Anthropics.

According to the *Collins English Dictionary* my parents bought me as a child – the one that mystified me as a gift one Christmas – the word 'anthropic' means 'of or relating to human beings'. In physics, the anthropic principle relates the fundamental laws to human existence or, more generally, the existence of complex and intelligent life. In the context of an unexpected universe, this provides an alternative to naturalness: some of the small numbers we find in nature are said to be there so that life can prosper, not because there is some mysterious symmetry or fancy new physics.

It's the science of life and death, and of a multiverse. But there are those who say it isn't even science.

The basic idea goes back to 1973, when Australian physicist Brandon Carter challenged the Copernican lore. Five hundred years earlier, Copernicus had humbly declared that we were nothing special, that our place in the universe wasn't privileged. Carter thought otherwise. It seemed as if the physical laws were perfectly tuned so that, once the symphony had begun, intelligent life could evolve. Steven Weinberg eventually showed how this logic could be applied to the cosmological constant, but others have applied it to other puzzles, not least the number of spatial dimensions or the unexpectedly low mass of the Higgs boson.

As we saw at the beginning of this chapter, the odds of a cosmological constant like ours are more than a googol to one. If the National Lottery offered similar odds, you probably wouldn't bother buying a ticket. But suppose you were determined to win – that your life depended on it – what would you do then? There is only one way to give yourself half a chance – you need to buy an awful lot of tickets. In the lottery of the cosmological constant, each ticket is a universe with a different overall vacuum energy. Nature can beat the odds by buying a multiverse of tickets, one for every possible universe with every possible cosmological constant. Most of these universes are too heavy, filled with too much vacuum energy for complex life to evolve, but some of them are more than a googol times lighter, just like ours. To enter into one of these lighter worlds, you need to get hold of a golden ticket. It is only here, in this privileged corner of the multiverse, where we might find great art or literature, or where science will dare to blossom and intelligent creatures start to ask questions about the cosmological constant.

But nature also needs somewhere to buy the tickets, golden or otherwise. This is where string theory is supposed to come in. As we will see at the end of the next chapter, string theory may offer us a multiverse, a landscape of different possible universes. Thanks to the sorcery of quantum mechanics, we can also find ourselves in one universe and then jump spontaneously into another. This is how nature works its way through all the tickets in its collection. Chances are the first ticket will reveal a universe with an enormous cosmological constant; so will the second and the third, and many, many more. Nature will skip through many of these at random, but what will they be like? In universes this heavy, would Lionel Messi be able to play football? Would The Beatles conquer America? Would dinosaurs still rule the Earth? In every case, the answer is a resounding no. To find one of the golden tickets, nature must skip into a universe with a *tiny* cosmological constant.

It's all because we are stardust. This is true of you; it's true of Lionel Messi; it's even true of a triceratops. Everything that makes us what we are and the planet on which we live was synthesized inside a star. But in order to evolve complex life, we don't just need stars, we also need galaxies. Without galaxies binding groups of stars together, heavy elements released by supernova explosions would disappear into empty space. Galaxies ensure that this debris is sometimes gathered together, occasionally making planets laden full of all the right ingredients for complex life to evolve. The golden ticket to life is a ticket to a universe with galaxies.

Weinberg realized that too much vacuum energy was a problem for galaxies. He noticed that if the cosmological constant was large and positive, it would force the universe to accelerate early. Stars would no longer have enough time to gather together to form the galaxies we need, before being pushed violently apart by the expansion of space. Contrast this with what happens when the cosmological constant is large and *negative*. Then there is no acceleration, but there is something much worse. Whenever the universe begins to feel a negative cosmological constant, it sets about halting the expansion. Space begins to contract and the universe ends in an apocalyptic crunch.

An updated version of Weinberg's calculations finds that galaxies will emerge only if the cosmological constant is no more than a few thousand times the value we see in our universe. These are the golden tickets we've been talking about. They admit entry to a bespoke corner of the multiverse where galaxies can exist and where life can evolve. The rest of the multiverse is barren. The trick with anthropics is to demand the existence of complex life, of creatures like The Beatles, or Messi, or even Zel'dovich, some of whom will ask difficult questions about the universe in which we live. But the minute we do this, we narrow the odds of our world. We no longer need to worry about those corners of the multiverse where the cosmological constant is too large. We are only interested in the golden tickets, in comparing universes where complex life can prosper.

We can ask again: what is the typical value for the cosmological constant? Because we are restricting attention to the golden tickets, the cosmological constant doesn't scan over such a large range of values. In fact, it can't be more than a few thousand times the value we see in our own universe. By applying the anthropic principle – by setting the scene for complex life – we have drastically reduced the allowed range of values for the cosmological constant. Our universe is no longer a googol-to-one outsider. We know it has a golden ticket – it has complex life – so the odds of finding the right cosmological constant are a few thousand to one. That's quite an improvement.

Anthropics may be clever, perhaps even a little sexy with its multiverse of different worlds, but it's divisive. Many of its critics worry that it strays too far from the borders of science, that it cannot be falsified, not even in principle. This is perhaps unfair. In 1997, Weinberg made a prediction. He and his collaborators[4] argued that if the vacuum energy were less than approximately 60 per cent of the total energy budget in our universe, anthropic arguments could not explain why it was so small. This was crucial to the paper being published. The editor of the *Astrophysical Journal* had an aversion to anthropics – he was persuaded to publish only because the paper offered a route to abandoning the idea altogether. The following year, the supernova teams led by Adam Riess and Saul Perlmutter announced the evidence for cosmic acceleration. We now know that the cosmological constant makes up about 70 per cent of the cosmic energy budget. Weinberg's prediction had come true. He tested anthropics, and anthropics passed the test.

The problem with anthropics, as with so many things, is that we are often biased by our own experience. Whenever we ask questions about life we look to our own surroundings and are heavily influenced by the diversity of our astonishing planet. But the moment we do that, we're compromised. I once asked a biologist if he thought alien life would also be based on DNA. He didn't know. How could he? He had never dissected an alien from another planet or, indeed, from another universe. The criteria we use to apply the anthropic principle and the existence of intelligent life are often littered with educated guesswork, and it's difficult to really know if the guesses are right.

And then there is the multiverse itself. Does it exist? We have no proof that it does, be it experimental or mathematical. String theory looks as if it predicts one, but we know very little about its structure. A key feature of anthropics is the ability to randomly jump from one universe to another. This might be possible with some quantum sorcery, but what if there are barriers in the multiverse that discourage this, or prevent it altogether? Without that detailed knowledge of the multiverse, there isn't an awful lot we can say that isn't weighed down by caveats and assumptions.

The theory of anthropics is the theory of life, the quest to understand the ultra-fine balance that exists in nature and allowed you and me to be born on a rocky planet, in the habitable zone around a medium-sized star. But it remains a theory of many unknowns, perhaps even unknowables. Should we really abandon naturalness for something this loose? My instinct says no. Naturalness is an appreciation of the beauty and elegance of nature. It is the search for its symmetry. It was symmetry that gave the photon a vanishing mass, so that light may travel at the speed of light. It was symmetry that prevented the electron from getting so heavy it destabilized the atom. But what is the symmetry protecting our universe from the energy of empty space? What is the beautiful new physics that tames the cosmological constant?

THE GHOST OF SIR ISAAC NEWTON

I had to stoop when I walked into the building. The ceilings were low, crossed by imposing wooden beams, and there were carvings on the wall that had been put there to keep the witches away. I was in Woolsthorpe Manor, an old farmstead steeped in history set deep inside the Lincolnshire countryside. It was here, in the early hours of Christmas Day 1642, that Hannah Newton gave birth to her eldest son, Isaac. A boy who would be king of science. A boy that Hannah said was small enough to fit inside a quart mug.

I'd gone to Woolsthorpe with a colleague from the University of California, in search of some inspiration. For two twenty-first-century physicists it doesn't get much better than this, and we hoped the ghost of Newton would be our invisible guide as we bounced ideas and equations to and fro in the shade of the apple trees that still grow in the manor's orchard.

It almost worked.

By the time we were thrown out of the manor at closing time we had formulated an exciting (and terrifying) new idea that connected the cosmological constant problem to an impending apocalypse. We weren't quite ready to go home, so we headed for the nearest pub, the White Lion, in the nearby village of Colsterworth. The pub is fairly traditional, with rugged stone walls and a wood-panelled bar, and it overlooks the Saxon church where Newton was baptized. As I handed my friend a pint of lager, he was scribbling some more equations on the back of a napkin. I took issue with him over some of the details and, as we argued, I noticed the curious glances from a group of bearded construction workers sitting at the table next to us.

'What are you two doing?'

The accent was local – Lincolnshire, strong and countrified. I was about to make up an answer that would make us seem a little less *geeky*, a little less like the over-indulged academics we undoubtedly were. But I was too slow. The American professor, less familiar with the unwritten rules of English pubs, immediately replied:

'We're working out when the universe will end.'

I should not have worried. For the next hour or so we explained our ideas to our new friends from the pub and they were fascinated. We talked about how the established view of the universe made no sense, how the vacuum of space should be a bubbling broth of quantum excitations, blasting the universe apart so violently that stars, planets and humans could never even exist. We claimed that we had an idea to resolve this conundrum but that it came at a cost: the universe would have to end, and soon.

Their looks of alarm were understandable. Of course we meant 'soon' in cosmological terms. Our friends were suitably relieved: a few tens of billions of years was easily enough time to get another round in.

The ideas we were bouncing around Woolsthorpe on that warm summer's day were inspired by a very simple observation: the cosmological constant is, you know, *constant*. It seems a bit obvious, but this really is what makes the cosmological constant special. It's what makes it different from planets and stars and everything else that affects the force of gravity.

Let's compare it to a planet. Like a cosmological constant, a planet will affect the gravitational field, but it does so quite differently. The mass of a planet isn't uniformly spread but gathered together in a small region of space and time. This means you have gradients in its profile – regions where the mass density begins to fall off. But the cosmological constant is different. It is, as far as we can tell, *constant*. In our corner of the universe and in our particular time, the underlying vacuum energy is unchanging. There are no gradients.

In Einstein's General Theory of Relativity, we know that all forms of energy gravitate – nothing is exempt. Spacetime will be bent by planets and stars, by humans and sentient bodies of alien gas. It will also be bent by vacuum energy. What we wanted to do was develop a new theory of gravity that treated the cosmological constant a bit differently. The planets and stars should gravitate just as Einstein said they would. So should you and I. But the underlying reservoir of vacuum energy – the constant – that shouldn't gravitate at all.

Our theory became known as *vacuum energy sequestering*. To sequester something is to isolate it or hide it away somewhere. The theory is very similar to Einstein's theory of gravity, but it comes equipped with a mechanism for hiding away the large vacuum energy predicted by quantum mechanics. To understand how it works, you should think about the way in which your fridge stays cold. The fridge has a thermostat set to some particular temperature, probably around four degrees Celsius. If the temperature inside rises above four degrees, the thermostat triggers an external cooling mechanism – it turns on the compressor and the refrigerant starts to circulate through the system. When the fridge has cooled back down, the thermostat switches the compressor off and the cooling stops. In vacuum energy sequestering, the universe also has a thermostat, but in this case it measures the *average* temperature of the universe across *all of space and all of time*.

Now imagine a universe with an overwhelming large vacuum energy, such as a googol gigajoules of energy in every litre of empty space. In General Relativity, this energy will bend and crush the universe into oblivion, raising its temperature to almost a billion trillion trillion degrees Celsius. But in vacuum energy sequestering, there is also the thermostat. In principle, it can be set to any value we choose, so we set it to within a whisker of absolute zero. In the presence of this enormous vacuum energy, the thermostat triggers an external cooling mechanism to lower the energy and bring the average temperature down to the desired value. Because it's an external mechanism – in this case, external to spacetime – it doesn't distinguish between one spacetime point over another. It doesn't distinguish between today and tomorrow, or between

America and Andromeda. It lowers the energy by an equal amount *at all points in space and time*. In other words, it lowers the base line, the underlying reservoir of vacuum energy. Other sources of energy like stars, planets and little green men are not affected by this change. Only vacuum energy is sequestered.

Protected by this thermostat, it's almost as if the universe has an element of precognition. Whatever the vacuum energy, it knows from the very beginning that it will survive. The thermostat ensures it will grow old, large and desolate, so much so that humans are able to evolve. You might think it sounds a little acausal, perhaps even a little like fate. Do you believe in fate? Most scientists would say that they don't, but what would happen if they crossed the event horizon of Pōwehi or any other black hole, as it sits on a throne at the centre of a galaxy? Would they not be destined to end their days in infinite torment, at one with the black hole singularity? The truth is their fate would be sealed the moment they crossed the event horizon, but that doesn't mean there is anything physically inconsistent about it. Causal paradoxes emerge only when time is caught in a loop, as in the story of a time traveller who goes back and eliminates their parents before they were conceived. But there is no obvious mechanism for anything like that to happen in our theory. There are no paradoxes. It's just that the universe has a destiny. Thanks to the thermostat, it knows it has to grow old and large.

This connection between the cosmological constant and cosmic precognition isn't new. It had been suggested decades earlier by a number of thinkers, most notably, Sidney Coleman. Coleman was the physicist's physicist, a student of Gell-Mann who built a formidable reputation within the academic community but remained strangely unknown to the outside world. What my American colleague and I did was build his idea into a simple working model.

But is it right?

I honestly can't say. What I can say is that it's not obviously wrong – and that's already an achievement in a field as mature as ours. We've been developing the idea for eight years now. I always know how long it's been because my daughter was born just as we released the first paper. Of course, I didn't time it like that deliberately – she wasn't supposed to arrive for another two months. But as my daughter grows up, our model continues to survive. It has not been ruled out by any observation, nor has it fallen victim to any mathematical inconsistency or catastrophic instability.

And what of the apocalypse? Didn't we tell our friends in the pub that it was imminent, at least in cosmological terms? For a while, we thought that it was. In our earliest models, it was the price we had to pay for conquering the cosmological constant. It made for a good conversation and it gave us a prediction, albeit an alarming one. But our models matured over time and, in the end, we realized that the apocalypse didn't always have to happen. Maybe one day I'll return to the White Lion and assure my friends that all is now well. If our latest models are correct, we can look forward to a longer cosmic future and still do away with the cosmological constant.

I said at the beginning of this chapter that physicists were embarrassed by the little numbers: by the cosmological constant; by the Higgs; by our desperately unexpected universe. But maybe we shouldn't be embarrassed. Maybe we should be celebrating. After all, the slimline Higgs and the slimline cosmological constant are trying to tell us something important about the fabric of our physical world. What could it be? What is the fundamental physics that drives them to such tiny values? Is it some unknown symmetry? Is it precognition, as in vacuum energy sequestering? Is it the existence of life itself, as in anthropics? I cannot say. What I can say is that these tiny little numbers are the portal of discovery. One day, we will figure out what they are trying to tell us, through the power of our mathematics, by pushing and pulling on the consistency of our ideas, and through the power of our experiments, by peering further and further into an unexpected world.

Infinity

Infinity

THE INFINITE GODS

Georg Cantor was much thinner than he used to be, his coat hanging heavily off his feeble body. His face was expressionless. There had been a time when he was a vibrant and imposing figure, invigorated by his own intellect and the pursuit of his mathematical dream. But there is no evidence of this in his last surviving photograph, taken in his home town of Halle in 1917. By then, the Great War had been raging for three long years and the German people were starving. The harvests had failed and the Allied warships were blocking the supply of food into the country. Some Germans were able to supplement their rations through farming or the black market. But not Cantor. He had been committed to a psychiatric clinic – the Nervenklinik in Halle – after suffering from manic depression. With food rations in German institutions less than half their normal amount, and mortality rates doubled, he wrote continuously to his wife, begging to return home. She couldn't grant him his wish. On 6 January 1918, his body weakened through malnutrition, Georg Cantor died of a heart attack.

Cantor's later life had been terrorized by mental illness, personal tragedy and professional exhaustion. But despite the lows that he had endured, he had also risen higher than anyone. He had dared to imagine the unimaginable, reaching into the heavens to look upon the celestial numbers – the infinities. Cantor did not just see infinity at the edge of the finite realm, but higher infinities, far beyond a terrestrial understanding. Thanks to his ideas, we now know that there are some infinities that are so big they are mathematically *inaccessible* to other, smaller infinities. In other words, there are realms of infinity beyond the realms of infinity.

More often than not, infinity is represented by a drunken number eight, ∞, lying on its side after a few too many tequilas. The symbol was introduced in 1655 by the Englishman John Wallis and is sometimes called the lemniscate, meaning 'ribbons'. But this particular infinity is not a number – it represents a *limit*, the idea of carrying on for ever, *ad infinitum*, beyond what you can ever hope to reach. But, as Cantor showed, infinite numbers do exist, and they are infinitely many. They are as tangible as five or forty-two, or even a googol. It's just that they don't exist in the finite realm – they are *transfinite*. They are the monstrous alephs and the mighty omegas, and there is even one called the Yeti.

Let us begin with some questions.

Did you know that there are as many even numbers as there are whole numbers?
Did you know that there are as many real numbers between zero and one as there are between zero and TREE(3)?
And did you know that there are as many points on the circumference of a circle as there are in its interior?

Things are rarely intuitive when infinity is involved. This is certainly true of the Hilbert Hotel, named after the great German mathematician David Hilbert, who conjured up the idea more than a century ago. The Hilbert Hotel has an infinite number of rooms, which means that even when it's full, the manager can take on as many new guests as he likes. To understand how he does this, we label the rooms by number: one, two, three, four, and so on, *ad infinitum*. When a new guest arrives, all the manager has to do is shuffle everyone along one: the family in room one move to room two, the couple in room two move to room three, the businessman in room three moves to room four, and so on. Thanks to the infinity, the process never breaks down. After everyone has shuffled along, the new arrival can settle into room one, which has just been vacated at the start of the chain. Even when he's faced with an infinite number of new guests, the manager doesn't panic. He simply moves everyone into a new room with twice the numerical value. All the existing guests now fill up the even-numbered rooms, while the new ones fill up the odd. There is always room at the Hilbert Hotel.

By his own admission, David Hilbert was a 'dull and silly boy' who did not initially impress at school, but he would grow up to become one of recent history's most influential thinkers. His work formed the foundation of much of modern mathematics *and* physics, from logic and proof theory to relativity and quantum mechanics. But perhaps he is best known for the list of twenty-three unsolved mathematical problems he published in 1900 which has had a profound impact on research over the last century. The first of these – the continuum hypothesis – is a problem of infinity, originally posed by Cantor. To this day, only eight of Hilbert's problems have solutions that are fully accepted by the mathematical community. As we will see later on, the continuum hypothesis isn't one of them.

The first recorded mention of infinity dates all the way back to the sixth century BC, to ancient Greece and the philosophical works of Anaximander. Anaximander was a master of the Milesian school, where he probably taught Pythagoras. Although most of his writings have been lost over the years, in the few surviving fragments, he speaks of infinity as *apeiron*. Taken literally, it translates as indefinite, boundless, or without limit. Anaximander was trying to understand the origin of all things. He imagined apeiron as the endless and inexhaustible soup from which everything is born, and to which it returns when it is finally destroyed. To an ancient Greek, it did not conjure up an image of beauty but of chaos. It was not heaven but the abyss.

Infinity and its infinitesimal cousin are at the heart of the paradoxes of Zeno of Elea. You may remember Zeno as the philosopher who conspired against the tyrannical rule of Nearchus. He was captured, tortured and eventually killed, after biting off a piece of the tyrant he had been desperate to overthrow. In chapter 'Zero' we told Zeno's paradox of Achilles and the tortoise, where the fleet-footed warrior is unable to run past the slow-moving reptile. In another paradox, the so-called 'dichotomy', Zeno asks a very simple question: how is it that you ever cross a room? At first glance, the question seems absurd, but Zeno conjured up an argument that challenged our everyday illusions. Consider where you are sitting reading this book. To get out of the room, you must first reach the halfway point between you and the door. But to get to the halfway point, you must first reach the quarter point; and to get there you must first reach the point one eighth of the way across. You can continue this sequence *ad infinitum*, until, like Zeno, you start to believe that motion is impossible.

This paradox demonstrates the subtle difference between an infinitesimal and zero. Zeno's trickery generated a sequence of rational numbers

$$\frac{1}{2}, \frac{1}{4}, \frac{1}{8}, \frac{1}{16} \cdots$$

Take any positive number you want, no matter how small; if we move far enough along Zeno's sequence, we can reach a number that is even smaller in a finite number of steps. But we can never actually reach zero, as Zeno would have us believe. Zero is the *limit* of the sequence, but it is not a part of it. As Aristotle would muse a century later, we can understand the *potential* to reach an infinity of steps, but we can never actually reach it. He believed that infinity was something you could hold in your mind but never in your hand. According to Aristotle and his followers, *potential* infinity was real, *actual* infinity was not.

The truth is the ancient Greeks had little appetite for *apeiron*. When Plato imagined the ultimate form of Good, he declared it to be finite and definite, never to be stained by the chaos of infinity. But as the Greeks began to lose their intellectual dominance, infinity began to rise. At the beginning of the third century AD, Plotinus, a Roman by birth, connected infinity to a supreme entity he described as *the One*. The One was understood to be beyond division and multiplication, a divine infinity that existed without limit. It is an idea which continued to resonate two centuries later with St Augustine's thoughts on the Christian God. By this time, Rome's power had crumbled, and many blamed its conversion to the new Christian religion. In response to this, Augustine was commissioned to write a series of books promoting Christianity and arguing in favour of its superiority over the old Roman ideology. It was in those books that he touched the infinite, inferring its existence in the mind of God. Augustine realized that numbers had to exist without limit, because if we declared there to be a largest number, we could always add one more. As there could never be a number of which God was ignorant, He must know all the numbers. He must be capable of infinite thought.

A connection between God and infinity can be found in many other religious contexts. For example, in Jewish mysticism, the Kabbalists spoke of the ten *Sefirots* and the underlying *Ein Sof*. While each of the Sefirots represented a different aspect of the divine body, the Ein Sof was something greater, an infinite god beyond description and comprehension. Similarly, in Hinduism, the god Vishnu is sometimes called *Ananta*, a Sanskrit word meaning endless or limitless. It can also mean infinity.

By the thirteenth century, Aristotle's ancient ideas, including his rejection of actual infinity, were beginning to re-emerge in the Western world. As a result, most medieval thinkers were reluctant to go as far as Augustine and accept God's ability to create infinities beyond His own existence. The most notable of these was St Thomas Aquinas, who argued that these limitations placed no restrictions on the power of God. His point was that additional infinities could not exist in actuality, as Aristotle had claimed, so it would be logically inconsistent for God to create them. Despite His unlimited power, God could not make something infinite, just as he could not make something unmade. The argument is superficially elegant, but on closer inspection we see that it's circular. It begins and ends with the same idea: that only finite things are allowed to exist.

As theology gave way to modern scientific ideas, few had the appetite to challenge infinity. Many Renaissance mathematicians tried to exploit the potential for infinity, in line with Aristotle, but they didn't dare touch it. They were content to *approach* infinity by taking bigger and bigger numbers, but they would never ask about infinity itself.

But Galileo was different.

He had already upset the establishment. In his *Dialogue on the Two Principal World Systems*, Galileo had taken on the Catholic Church, arguing in favour of the Copernican world view, with the Sun at the centre and the Earth a peripheral bystander. His book was presented as a conversation between three men: the academic Salviati who would try to convince his friends of this heliocentric model; an intelligent layman, Sagredo; and the dimwitted Simplicio, a backward and traditional character who many took to represent the Pope. Led by the pontiff's nephew, Cardinal Francesco Barberini, the Church responded swiftly to the offence. Galileo was ordered to present himself in Rome and stand trial for heresy.

Luckily, the great scientist had powerful friends. The grand duke of Tuscany looked to intervene on his behalf, and he was even offered asylum in the Republic of Venice. Perhaps through arrogance or naivety, Galileo declined all these offers and chose to defend himself before the Inquisition. He believed he had been given permission to publish his ideas by the late Cardinal Bellarmine, and even had a letter to prove it. Unfortunately, the details didn't quite match the copy of the letter that was held in the Vatican. The Inquisition soon found him guilty and demanded he recant his work or face torture and death. As Galileo knelt before them, renouncing the Copernican view, it is said that he muttered in defiance, *E pur is muove*. And yet it moves.

Galileo spent the rest of his life under house arrest, during which time he wrote his masterpiece: *Discourses and Mathematical Demonstrations Relating to Two New Sciences*. In this work he developed his ideas on motion, forming the bedrock upon which others, from Newton to Einstein, would eventually build the tower of modern physics. It was in this final book that Galileo dared to touch infinity. As ever, he presented it as a conversation between his three main characters, although, with the Church watching, Simplicio was a little smarter than before.

In Galileo's story, Salviati invites his two friends to think about the infinite family of square numbers. Simplicio, tied up in the knots of Aristotle, is unhappy with Salviati's infinite recklessness. But Sagredo encourages him to go on and, soon enough, Salviati encounters a paradox. If you take all the whole numbers between zero and fifteen, you see that only four of them are squares, specifically zero, one, four and nine. Likewise, if you take the whole numbers between zero and ninety-nine, you find that only ten of them are squares. If we extrapolate to infinity, we're tempted to say that there are many more whole numbers than square numbers. After all, every square number is also a whole number, but the converse is not true.

Except now we're dealing with infinity, and infinity bites.

Salviati realizes that every square can be labelled by its square root. For example, $0 \to 0$, $1 \to 1$, $4 \to 2$, $9 \to 3$, and so on. With this relabelling we can turn the family of squares into the family of natural numbers: 0, 1, 2, 3, and so on. The point is that the matching between the two families is *one to one*: for every square there is a corresponding natural number given by the square root, and for every natural number there is a corresponding square. This must mean that the two families have exactly the same size! Although this is actually correct, Salviati is reluctant to draw this conclusion too quickly and chooses instead to declare an infinite ambiguity. He decides that any notion of comparison – greater, less or equal – seems not to apply to infinite quantities. However, such comparisons can be made as long as we stick to a particular set of rules. We declare equality whenever there is this one-to-one matching between families, or 'sets', as they would later become known. This might seem counterintuitive when an infinite family is matched to some, but not all, of its individual parts, but it doesn't lead to any mathematical breakdown. It's why we can say there are exactly as many natural numbers as there are even numbers, or squares, or powers of TREE(3).

After Galileo had dabbled in the infinite occult, it would be another two hundred years before anyone was brave enough – or foolish enough – to follow. Warnings to stay away from these occult practices came from the highest of authorities, the so-called *Princeps Mathematicorum* (the foremost mathematician), Carl Friedrich Gauss. In his letter to his fellow German Heinrich Schumacher in 1831, Gauss had warned, 'the use of an infinite quantity as a completed one ... is never permissible in mathematics. The infinite is only a *façon de parler*, where one is really speaking of limits to which certain ratios come as close as one likes while others are allowed to grow without restriction.' But a disgraced Catholic priest from the city of Prague was determined to think differently. His name was Bernard Bolzano.

Bolzano was the son of an Italian art dealer, a devout Roman Catholic also called Bernard who used his wealth to give generously to the poor, setting up an orphanage in his adopted city of Prague. These acts of compassion had a lasting influence on Bolzano, who would spend much of his adult life battling for fairness and greater equality. He would also tackle infinity.

By his own admission, Bolzano was a moody child, and sick, suffering impaired vision and intense headaches. At school, he was academically unexceptional and unpopular among his peers. He spent his childhood on the periphery, but where others might have grown inwards, the isolation seems to have equipped him with an independence of thought and the rare capacity to challenge accepted wisdom. As a young man, Bolzano studied for his doctorate in theology and, soon after, he was ordained as a Catholic priest. He quickly developed a reputation as a free-thinking Christian philosopher and was awarded the position of Chair of Religious Philosophy at Charles University in Prague at the age of just twenty-four. Bolzano never subscribed to Christian mysticism but justified his faith on moral grounds, in helping to achieve good in a society tainted with cruelty and hardship. At the time, Prague was heavily influenced by religious conservatism and in the years that followed he, like Galileo before, began to upset the establishment. Bolzano was preaching pacifism and a form of socialism to his students. It went largely unnoticed until Jakob Frint, a leading theologian who also served as confessor to the emperor in Vienna, encouraged Bolzano to use his new textbook as part of his course. Bolzano refused. In his opinion, the book was incomplete and too expensive for students. Frint was affronted and began to turn people against Bolzano, pointing out the radicalism of his sermons and his refusal to accept conservative Christian values. Bolzano enjoyed the support of his friend, the Archbishop of Prague, but the campaign against him continued. He stuck to his beliefs and continued to preach against war, private ownership and the bohemian establishment and eventually – inevitably – he was dismissed. Still in his early forties, he was pensioned off and asked to leave the university. Moving between the city of Prague and the surrounding countryside, he turned away from religion and towards mathematics – towards infinity.

He asked himself a simple question: if he were to hold infinity in his hand, what could it be? Gauss and others had declared it to be a boundless changeling, a variable quantity that grows and grows, without ever halting, without ever reaching its limit. But Bolzano rejected this: a variable quantity is not really a true quantity at all but the *idea* of a quantity. That was insufficient – it would be like saying you have x eggs in a basket, even after you've already counted them!

Bolzano recognized the family of natural numbers as the genuine article, an actual infinity he could use to quantify other infinities. He realized that *anything* he could match to these numbers, one to one, must also be an actual infinity. To make this more rigorous, he began to develop the idea of a set. A set is just a collection of things, like the 'Four Horsemen of the Apocalypse' or 'teams that play in the Premier League'. These are examples of finite sets: there are four Horsemen of the Apocalypse and twenty teams that play in the Premier League. But Bolzano was also willing to think of infinite sets – like the set of natural numbers, or the set of real numbers between zero and one. And he was convinced that these things actually existed. It didn't matter that you couldn't split them up and imagine all their individual parts. As Bolzano argued, it is completely reasonable to talk about the set of people who live in the city of Prague without having a mental picture of every single one of them. He applied similar logic to his infinite sets.

Confident in the existence of his infinite playground, Bolzano decided to play. Two centuries earlier, Galileo had discovered a paradox, demonstrating a one-to-one match between the set of natural numbers and the set of squares. But Bolzano went further, leaping into the continuum and finding his own paradox: he showed that there are as many real numbers between zero and one as there are between zero and two. Roughly speaking, this is how he did it. He started off with the smaller interval, between zero and one, and he doubled every single number. For example, $0 \rightarrow 0$, $0.25 \rightarrow 0.5$, $0.75 \rightarrow 1.5$, $1 \rightarrow 2$, and so on. This gave him a new set of numbers starting at zero and ending at two, filling up the space in between. He also realized that he could reverse the process, passing from the larger interval to the smaller one by halving every number. This might all seem really obvious, but what Bolzano had created was a simple one-to-one matching between our two continuous sets, just as Galileo had with the set of natural numbers and squares. Using the logic of one-to-one matching, we can argue that there are as many real numbers between zero and one as there are between zero and two, or zero and TREE(3), or even a googol and Graham's number.

Galileo had stopped short of saying that his infinite sets were equally numerous, even though it is consistent to do so. Bolzano was similarly cautious with his continua. Even though the one-to-one matching suggested that there were as many numbers between zero and one as between zero and two, he couldn't quite believe it. It is this reluctance that prevented him from going any further. Bolzano died before any-one really paid any attention to his work. In the meantime, other important mathematicians entered the infinite fray and, by the mid-nineteenth century, the stage was finally set. Galileo and Bolzano had had the courage to touch infinity, but it was Georg Cantor who reached high into the heavens. He dragged himself up and walked among the infinite, as no one had ever imagined possible.

THE ALEPH AND THE OMEGA

'The time will come when these things which are now hidden from you will be brought into the light.'

The motto appears at the beginning of one of Cantor's final publica-tions, in 1895. It is taken from the Bible, first Corinthians, and it betrays Cantor's belief in the divinity of his task. To Cantor, it was God who had led him to this infinite paradise, this infinite hell. It was God who had communicated through him, who had presented him with the aleph and the omega. There are even echoes of Revelation: 'I am the Alpha and the Omega. The First and the Last. The Beginning and the End.'

It is easy to dismiss this as religious delirium, and perhaps it was, but Cantor was inspired by his religious quest. As others around him attacked him for his infinite recklessness, as they called him 'a charla-tan' and 'a corrupter of youth', Cantor stood firm, emboldened by his faith. He had the courage to take on infinity, and he won. But he also lost. Cantor was overcome by the magnitude of his quest and plunged into a deep depression from which he would never fully escape.

Cantor began by accepting something that Galileo and Bolzano would never fully accept: if two sets could be matched, one to one, they must have exactly the same size. Of course, for finite sets, this isn't con-troversial. Take, for example, the Four Horsemen of the Apocalypse:

{Death, Famine, Pestilence, War}

and another set, famously known as The Beatles:

{John, Paul, George, Ringo}.

The two sets are easily matched up in one-to-one correspondence: Death could go with John, Famine with Paul, Pestilence with George and War with Ringo. There isn't anything special about how we do the matching – we could just as well have matched Death with Paul and Famine with John. The important thing is that every one of the horsemen is paired up with a different Beatle, and vice versa, and no one is left out. This all works out just fine because The Beatles and the Four Horsemen of the Apocalypse clearly correspond to sets of the same size. Things get a little more unsettling with infinite sets, as we have already seen. The set of square numbers is easily matched, one to one, with the set of whole numbers, even though it *appears* to be somewhat smaller. But Cantor understood that appearances can sometimes be deceptive, especially where infinity is involved.

Mathematics is a game where you make your own rules and, as long as it doesn't run into any logical inconsistencies, you are always good to go. Cantor defined the size of a set through its *cardinality*. The Beatles and the Horsemen of the Apocalypse are sets of cardinality four because we can match them up, one to one, with the first four natural numbers $\{0, 1, 2, 3\}$. (Remember, most mathematicians prefer to start counting at zero.)

$$Death \leftrightarrow John \leftrightarrow 0$$
$$Famine \leftrightarrow Paul \leftrightarrow 1$$
$$Pestilence \leftrightarrow George \leftrightarrow 2$$
$$War \leftrightarrow Ringo \leftrightarrow 3$$

The set of teams in the Premier League has cardinality twenty because we can match it up, one to one, with the first twenty natural numbers $\{0, 1, 2, 3 \dots 18, 19\}$. So what about some of our infinite sets? Because of the one-to-one matching, Cantor realized that the set of all squares $\{0, 1, 4, 9 \dots\}$ must have the same cardinality as the full set of natural numbers $\{0, 1, 2, 3 \dots\}$.

But how many numbers are there? What is the cardinality of the set?

It's not four or twenty or even TREE(3). It must be something bigger, something more infinite. Cantor decided to call it aleph zero, written as $\aleph_0$, with the first letter of the Hebrew alphabet. The subscript zero is a hint that this will only be our first infinity – there will be many more. But for now, be patient. If this first infinity is *defined* to be the cardinality of the set of natural numbers, then, thanks to the one-to-one matching, it is also the cardinality of the squares, the evens, the multiples of Graham's number and the powers of TREE(3). In a remarkable display of mathematical trickery, Cantor also showed that $\aleph_0$ was the cardinality of the rationals – the numbers you can write as a fraction of whole numbers.

Let's see how he did it.

Cantor began his proof by writing out all the fractions in a systematic way:

$\frac{1}{1}$	$\frac{2}{1}$	$\frac{3}{1}$	$\frac{4}{1}$	$\frac{5}{1}$	$\cdots$
$\frac{1}{2}$	$\frac{2}{2}$	$\frac{3}{2}$	$\frac{4}{2}$	$\frac{5}{2}$	$\cdots$
$\frac{1}{3}$	$\frac{2}{3}$	$\frac{3}{3}$	$\frac{4}{3}$	$\frac{5}{3}$	$\cdots$
$\frac{1}{4}$	$\frac{2}{4}$	$\frac{3}{4}$	$\frac{4}{4}$	$\frac{5}{4}$	$\cdots$
$\frac{1}{5}$	$\frac{2}{5}$	$\frac{3}{5}$	$\frac{4}{5}$	$\frac{5}{5}$	$\cdots$
$\vdots$	$\vdots$	$\vdots$	$\vdots$	$\vdots$	$\ddots$

If this table were to go on for ever in all directions, it would include every single one of the rational numbers. Of course, there would be a lot of repetition, but we will handle that. The question is: can we match all the distinct entries in this table, one to one, with the set of whole numbers? You might try to do this by first counting along one of the rows, matching up the fractions with the whole numbers, one by one. For example, if you started on the second row, you would write the following:

$\frac{1}{1}$	$\frac{2}{1}$	$\frac{3}{1}$	$\frac{4}{1}$	$\frac{5}{1}$	$\cdots$
$\frac{1}{2} \rightarrow 0$	$\frac{2}{2} \rightarrow 1$	$\frac{3}{2} \rightarrow 2$	$\frac{4}{2} \rightarrow 3$	$\frac{5}{2} \rightarrow 4$	$\cdots$
$\frac{1}{3}$	$\frac{2}{3}$	$\frac{3}{3}$	$\frac{4}{3}$	$\frac{5}{3}$	$\cdots$
$\frac{1}{4}$	$\frac{2}{4}$	$\frac{3}{4}$	$\frac{4}{4}$	$\frac{5}{4}$	$\cdots$
$\frac{1}{5}$	$\frac{2}{5}$	$\frac{3}{5}$	$\frac{4}{5}$	$\frac{5}{5}$	$\cdots$
$\vdots$	$\vdots$	$\vdots$	$\vdots$	$\vdots$	$\ddots$

But this strategy is never going to work – you will never be able to move on to the next row before running out of fuel. Cantor, in contrast, had a much better idea. He decided to snake his way through the table along a line of growing diagonals, skipping past entries (in grey) that can be simplified:

$\frac{1}{1} \to 0$	$\frac{2}{1} \to 1$	$\frac{3}{1} \to 4$	$\frac{4}{1} \to 5$	$\frac{5}{1} \to 10$	$\cdots$
$\frac{1}{2} \to 2$	$\frac{2}{2}$	$\frac{3}{2} \to 6$	$\frac{4}{2}$	$\frac{5}{2}$	$\cdots$
$\frac{1}{3} \to 3$	$\frac{2}{3} \to 7$	$\frac{3}{3}$	$\frac{4}{3}$	$\frac{5}{3}$	$\cdots$
$\frac{1}{4} \to 8$	$\frac{2}{4}$	$\frac{3}{4}$	$\frac{4}{4}$	$\frac{5}{4}$	$\cdots$
$\frac{1}{5} \to 9$	$\frac{2}{5}$	$\frac{3}{5}$	$\frac{4}{5}$	$\frac{5}{5}$	$\cdots$
$\vdots$	$\vdots$	$\vdots$	$\vdots$	$\vdots$	$\ddots$

It really is wonderfully clever. Cantor's strategy will never break down and, by the time he has snaked his way through the entire table, every fraction has been matched to a natural number. The cardinality of the rationals is proven to be $\aleph_0$.

The cardinality of sets gives us a way to talk about numbers. Actually, what we're really talking about are *cardinal numbers* – we'll encounter another type of number in a moment. Cardinal numbers are a way to measure *how many* things you have. They include all the finite numbers, like 0, 1, 2, 3 and, of course, our first infinity $\aleph_0$. But can we go any higher? Can we have more than $\aleph_0$?

What about $\aleph_0 + 1$?

To figure out what this is, we take an infinite set of rubber ducks with an infinite number of patterns, one for each of the natural numbers:

Clearly, there are $\aleph_0$ of them. In order to reach $\aleph_0 + 1$ we add another rubber duck – let's say a white one. It doesn't matter where we put it, so we might as well put it at the start and shove all the other ducks along one:

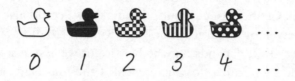

How many have we got now? Well, every duck is matched with a whole number, so it must be that there are $\aleph_0$. In other words, it turns out that $\aleph_0 + 1 = \aleph_0$. That's weird. What about $\aleph_0 + \aleph_0$? For this, we take two infinite sets of ducks, each of size $\aleph_0$, but this time we label one of them with the even numbers:

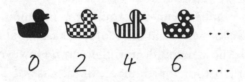

and the other with the odd:

When we bring them all together:

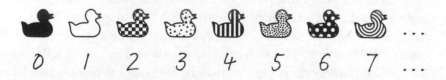

we immediately realize that $\aleph_0 + \aleph_0 = \aleph_0$. This is all a little strange. We're seeing things we just don't see with finite numbers. But why should we? We are now in the realm of the infinite.

I promised you more infinities, but it seems as if we're struggling to get past $\aleph_0$. To step beyond, we first need to restore some order. Up until now, we've been saying that our sets are arranged any which way but loose. For example, we said that The Beatles were given by {John, Paul, George, Ringo}, but presumably we could also have said they were given by {George, John, Paul, Ringo}. No difference, right? Not necessarily. It depends on whether or not we assign some sort of order, some significance to where each of the musicians appears in the set. In the second version of the set, {George, John, Paul, Ringo}, The Beatles have been arranged in alphabetical order. Even in the first version, you might argue that they've been arranged in order of talent – deeply controversial, I know (especially to my wife, who claims that Ringo is best because he was the voice of Thomas the Tank Engine).

The moment we start thinking about order, we change the rules of the game and numbers can take on extra meaning. Consider the number four. We know we can think of it as a cardinal number telling us, for example, how many Beatles there are. However, we can also think of it as a label for fourth place. In the case of The Beatles, we might associate it directly with Ringo because he appears fourth in the alphabet. When we do this we're thinking of four as an *ordinal number*, where we care about its position in the conveyor belt of natural numbers. The difference between ordinals and cardinals isn't so important until you step outside the finite realm and begin to play with infinity.

A convenient way to define an ordinal number is, of course, in terms of sets. We touched on this in chapter 'Zero'. We begin by thinking of zero as the empty set, one is the set containing 0, two is the set containing 0 and 1, three is the set {0, 1, 2}, and so on. In fact, every ordinal is defined as the set of ordinals that have gone before, $n + 1 = \{0, 1, 2, 3 \ldots n\}$. This is all very nice, but how will it ever take us to infinity and beyond? Well, to reach infinity, all we have to do is define the ordinal number that is one step along from all the finite ordinals. To do this, Cantor needed a new name and a new symbol. With the cardinal infinity already known as the aleph, he took inspiration from the divinity of his quest: 'I am the Alpha and the Omega.'

Omega, written symbolically as ω, would become the first of his ordinal infinities. If every finite ordinal was defined according to the rule, n + 1 ={0, 1, 2, 3 ... n}, then it was natural to define ω as the never-ending limit of this:

$$\omega = \{0, \ 1, \ 2, \ 3 \ldots\}.$$

In other words, the first of our ordinal infinities is nothing more than the set of natural numbers!

Let's take it higher.

What comes after ω? $\omega + 1$, of course. If we follow our rules, this is also defined as the set of ordinals that has gone before – in other words, it's the set of natural numbers with an omega cherry on top:

$$\omega + 1 = \{0, \ 1, \ 2, \ 3 \ldots; \omega\}.$$

We've used a semicolon to represent the boundary between the never-ending list of finite contributions, 0, 1, 2, 3. . . and the trans-finite contribution from ω. But this is just notation and not especially important. What is important is the fact that $\omega + 1$ is *not* the same as ω. This is because ordinals care about order. To understand this better, let's return to our ducks, only now we imagine they are real ducks, and they are in a race:

The black duck finishes first and is a bit annoyed because he is awarded with a zero. However, zero is the first of our natural numbers, so he shouldn't really complain. The checquered duck finishes second, and gets the second natural number (1), the stripy duck finishes third and gets the third natural number (2), and so on. The ellipses indicate that there are an infinite number of ducks in the race and each of them is awarded with a natural number. But suppose there is a second race with one more participant – the white duck. He is rather slow and crosses the line after all the others have finished. The picture looks a little like this:

When we added the white duck in our previous discussion, we didn't care about the order, so we just planted it alongside the black duck and shuffled all the others along one. This was how we showed that $\aleph_0 + 1 = \aleph_0$. But now we do care about the order – after all, it's a race! The white duck finished last, behind everybody else, so it isn't allowed to just blag its way to the front. What number should we assign to it? It can't be any of the natural numbers because they've all been used up, so it must be the next one on the list, which is ω. Because the order matters, it is clear that our two races are very different. The set of natural numbers is not the same as the set of natural numbers with an omega cherry on top, or in other words, $\omega + 1$ is *not* the same as ω.

We can carry on climbing. After $\omega + 1$ comes $\omega + 2$, once again defined in terms of the ordinals that went before:

$$\omega + 2 = \{0, \ 1, \ 2, \ 3. \ . \ .; \omega, \omega + 1\}.$$

It appears as if we've dragged ourselves up to the heavens, reaching up to a new ladder: from $\omega + 2$ on to $\omega + 3$, and so on, until we find a new layer of heaven at $\omega + \omega$. This is usually written as $\omega \times 2$ and is defined as the set:

$$\omega \times 2 = \{0, \ 1, \ 2. \ . \ .; \omega, \omega + 1, \omega + 2. \ . \ .\}$$

We can carry on climbing, to higher and higher heavens, up to $\omega \times 3$ and $\omega \times 4$, and so on, until we get to the limit, $\omega \times \omega$, which most sensible people would simply call ω^2. We have now reached an infinity of layers of infinite heavens. But we can go on. Climbing from here in the same way as before, we reach up to ω^3 and then ω^4 until eventually we reach another limit, an exponentially high heaven, which we write as ω^{ω}.

Now let's turn on the boosters.

From ω^{ω} we can imagine climbing higher and higher and higher, to a power tower of ω that is ω levels high:

$$\left.\omega^{\omega^{\cdot^{\cdot^{\omega}}}}\right\} \omega \text{ levels high}$$

As we saw in the Chapter 'Graham's Number', we can write these towers more efficiently with our double arrows, $\omega \uparrow\uparrow \omega$. From there we move on to

$$\omega \uparrow\uparrow\uparrow \omega = \underbrace{\omega \uparrow\uparrow \big(\omega \uparrow\uparrow \big(\ldots \uparrow\uparrow \omega\big)\big)}_{\omega \text{ repetitions}}$$

and then $\omega \uparrow^4 \omega$, and so on, until we reach yet another gargantuan limit, $\omega \uparrow^{\omega} \omega$, a heavenly leviathan, towering God-like above all that has gone before.

Remember when you used to think Graham's number was big?

But we're still not finished yet.

The funny thing about $\omega + 1$ is that it's not actually any bigger than ω – it just comes after it. The size of the corresponding set $\omega + 1 = \{0, 1, 2, 3 \ldots; \omega\}$ is still $\aleph_0$. To prove this, you just have to match the elements of $\omega + 1 = \{0, 1, 2, 3 \ldots; \omega\}$ with the natural numbers. That's easy: you just match ω up with 0; 0 with 1; 1 with 2; 2 with 3; and so on. Likewise, climbing up to $\omega + 2$ or even $\omega \uparrow^{\omega} \omega$ takes us to higher infinities, higher up on the list, but not to *bigger* infinities. They all have the same cardinality: aleph zero.

And then it happens.

At a genuinely unimaginable height, Cantor showed that there must be a new type of ordinal, different from everything that has gone before. It is not obvious that such a thing should ever exist, but it does. Cantor showed that larger infinities were hidden in the *continuum*, in the set of every real number, including rational numbers that can be written as fractions, and irrational numbers like $\sqrt{2}$ or π that cannot be written this way.* He showed that the continuum is beyond our terrestrial ability to count – one, two, three, four … It is bigger than aleph zero.

Let's ask how many real numbers there are in the continuum between zero and one. Infinity, for sure, but is it aleph zero or is it really something larger? Here's how Cantor figured it out. We begin by assuming the continuum is countable and can therefore be matched up, one to one, with the natural numbers. This must mean we can write them all out in an infinite list of size $\aleph_0$. The order doesn't matter, so we just start listing all the numbers between zero and one at random:

0.12347348956792457 …
0.34579479867439087 …
0.73549874397493486 …
0.42784508734067383 …
0.54345689483459808 …
⋮

To prove that the continuum was bigger than $\aleph_0$, Cantor showed that this list could not capture everything. He began by highlighting each number along the diagonal:

0.12347348956792457 …
0.34579479867439087 …
0.73549874397493486 …
0.42784508734067383 …
0.54345689483459808 …
⋮

This pointed towards the number made up of all the diagonal entries, in this case 0.14585 … Cantor then created a new number by changing every single one of those entries by one. In the example, 0.14585 … is converted into the new number 0.25696 … This

---

* Irrationals like $\sqrt{2}$ are sometimes known as algebraic, because they are solutions to simple algebraic equations with integer powers of x multiplied by whole numbers (for example, $\sqrt{2}$ solves the simple equation $x^2 - 2 = 0$). Irrationals like π or e aren't even algebraic numbers – they are the so-called transcendentals.

number differs from the first number in the list at the first decimal point, from the second number at the second decimal point, from the third number at the third decimal point, and so on. In fact, it differs from all $\aleph_0$ of them! This proves that it is impossible to capture all the numbers of the continuum with a list of size $\aleph_0$. It follows that the continuum hides a larger infinity, just as Cantor had imagined.

Is there a way to systematically construct these larger infinities, to navigate our way beyond aleph zero? The answer is yes. We've already identified a huge tower of infinite ordinals each of size $\aleph_0$, from $\omega = \{0, 1, 2, 3 \ldots\}$ and $\omega + 1 = \{0, 1, 2, 3 \ldots; \omega\}$ to $\omega \uparrow^\omega \omega$, and even higher ordinals. These are sometimes called the countable infinities, because each of them is really a set of numbers that can be matched one to one with the natural numbers – the numbers we use to count. But what lies beyond the tower? What is the ordinal one step *beyond* the countable infinities? This is omega one, written symbolically as ω_1. By definition, it cannot be countable – it cannot be matched one to one with the natural numbers. This heavenly giant must have a new cardinality, a new size. That size is aleph one, $\aleph_1$. It isn't just a higher infinity. It's a bigger one.

As usual, ω_1 is defined as the set of ordinal numbers that preceded it. In other words, it's the full set of countable numbers, from the finite tiddlers to the largest of the countable infinities. But from ω_1 we can carry on climbing, up to $\omega_1 + 1$ and even higher. Once again, these aren't necessarily any bigger than ω_1 – they just come after it. $\omega_1 + 1$ also has size $\aleph_1$ because it can be matched up one to one with the set of countable numbers. And then there is another layer, an ordinal beyond those of size $\aleph_1$. This is ω_2, an even bigger number with a magnificent new size, $\aleph_2$.

I imagine you are becoming infinitely unsettled by all of this. After all, infinity was difficult to comprehend, and yet now we're dealing with infinities beyond the infinite, with monstrous alephs and mighty omegas. Here is a little table to help you gather your thoughts.

Ordinal number	Definition of terms of sets	Description	Cardinality/ size	
0	{ }	The empty set	0	
1	{0}	Set of one element	1	NATURAL NUMBERS
2	{0,1}	Set of two elements	2	
3	{0,1,2}	Set of three elements	3	
⋮				
ω	{0,1,2, …}	Set of all natural numbers	$\aleph_0$	
$\omega + 1$	{0,1,2, …; ω}	Set of all natural numbers with an ω cherry on top	$\aleph_0$	
⋮				
$\omega \times 2$	{0,1,2,…; $\omega, \omega + 1, \omega + 2, …$}	The limit of the set $\omega + n$ as n goes to infinity	$\aleph_0$	
⋮				COUNTABLE INFINITIES
ω^2		The limit of the set $\omega \times n$ as n goes to infinity	$\aleph_0$	
⋮				
ω^ω		The limit of the set ω^n as n goes to infinity	$\aleph_0$	
⋮				
⋮				
ω_1		Set of all natural numbers and all countable infinities	$\aleph_1$	
$\omega_1 + 1$		Set of all natural numbers, all countable infinities, with an ω_1 cherry on top	$\aleph_1$	INFINITIES OF SIZE $\aleph_1$
⋮				
⋮				
ω_2		Set of all natural numbers, countable infinities and infinities of site $\aleph_1$	$\aleph_2$	EVEN LARGER INFINITIES

The infinities of Georg Cantor.

As it happened, Cantor reached beyond $\aleph_2$, towards higher layers of infinity, towards new heavens and new gods. But at the time, there were few who believed in his heavenly quest. On the contrary, he was in hell – at least that was the view of a mathematician named Leopold Kronecker. In the mid-nineteenth century, Berlin was the centre of the mathematical world, and Kronecker was one of the university's most influential professors. He was brilliant but conservative. 'God made the integers,' he said, 'all else is the work of man.' He was appalled by irrational numbers. Of course, he understood the mathematics behind them, but he saw no place for them in the natural world. They were the 'work of man', a fantasy for self-indulgent charlatans – like Cantor. Kronecker had been a mentor to Cantor, his tutor and his friend. But when Cantor moved south from Berlin, to the University of Halle, he would break free from the conservatism of his teacher. He would reach beyond the integers, towards the continuum and the new layers of infinity that lurked within. And Kronecker didn't like that.

The two men were eventually at war with one another, and it soon became personal. Kronecker routinely insulted Cantor and obstructed publication of his work in respected journals. It didn't matter that Cantor's ideas were as robust as they were remarkable; Kronecker had the political edge. Cantor was a professor at a second-tier university, while Kronecker was part of the establishment in Berlin. More than anything else, the sense of injustice tore at Cantor. He knew that he deserved more, that his abilities warranted a professorship in Berlin but, with Kronecker's destructive influence, he also knew that could never happen.

As the attacks continued, Cantor became increasingly desperate. In a bizarre attempt to strike back, he applied for a professorship in Berlin. Although he knew he had no chance of success, he was sure he would offend Kronecker, and that was enough. He told the Swedish mathematician Gösta Mittag-Leffler, 'I knew precisely [that] Kronecker would flare up as if stung by a scorpion, and with his reserve troops would strike up such a howl that Berlin would think that it had been transported to the sandy deserts of Africa, with its lions, tigers, and hyenas.'

With his moods and explosive personality, Cantor didn't have many friends, but Mittag-Leffler was one of them. A year earlier, in 1882, Mittag-Leffler had founded the journal *Acta Mathematica*, offering Cantor a safe place to publish his work, away from Kronecker's scheming. When Kronecker found out what was happening, he saw a way to take his revenge on his former student. He wrote to the Swede asking if he could publish in his new journal. Cantor got wind of this and sensed another attack. If Kronecker was hoping to publish in *Acta*, it could only be to discredit his work. Cantor responded with a characteristic explosion, writing angrily to Mittag-Leffler, threatening not to send him any more papers. The relationship between Cantor and his friend was damaged, perhaps as Kronecker had anticipated. He had never intended to send a paper to the journal.

Within a year, Cantor had suffered his first nervous breakdown. It is thought he may well have had problems with mental illness even if he had otherwise led a quiet and ordinary life. But the truth is he didn't. His life was engulfed by the intensity of his work and his battles with Kronecker. Later, he was also overwhelmed by personal tragedy, suffering the loss of his youngest son, Rudolf, who died suddenly while Cantor was away in Leipzig delivering a lecture in 1899.

Cantor had reached to the heavens, to the infinite, and walked among the alephs and omegas. As a deeply religious man, he believed he was guided by God. He was certainly guided by numbers – by *all* the numbers – by the *continuum*. It was here, in this celestial realm, that Cantor first glimpsed beyond aleph zero. He saw that the continuum was a bigger brand of infinity – a bigger aleph – but which one was it? Was it $\aleph_1$, or something even bigger?

Whenever we have a set, be it the Four Horsemen of the Apocalypse or the set of natural numbers, we can talk about something called the power set. The power set is just the set of all subsets. For example, consider the set of the Three Musketeers, {Athos, Porthos, Aramis}. From here we could make a grand total of eight different subsets. These are the empty set:

{ }

sets with a single musketeer,

{Athos}
{Porthos}
{Aramis}

sets with two musketeers,

{Athos, Porthos}
{Porthos, Aramis}
{Aramis, Athos}

and, of course, the set with all three,

{Athos, Porthos, Aramis}.

Together, these eight different sets form the power set of the Three Musketeers. You might have noticed that the Three Musketeers is a set of size 3, whereas its power set is a much larger set of size $8 = 2^3$. This isn't a coincidence. Each set in the power set has Athos in or out, Porthos in or out, and Aramis in or out. That triggers $2 \times 2 \times 2$ possibilities. By the same logic, if the set of teams in the Premier League has size 20, then its power set has size 2^{20}.

The rule also applies to infinite sets. We know that the natural numbers is a set of size $\aleph_0$. What about its power set? That is just the set of subsets of the natural numbers, or in other words, the empty set:

{ }

the single number sets:

{0}
{1}
{2}
⋮

the two number sets:

{0, 1}
{0, 2}
{1, 2}
⋮

and so on. This power set is a set of size $2^{\aleph_0}$. That is unimaginably huge. As Cantor proved, it is certainly bigger than $\aleph_0$, and it just so happens to be the size of the continuum. To see this, imagine

writing a real number in binary format. This is just a bunch of
zeros and ones arranged in some particular order. For example,

$$\frac{5}{8} = 1 \times \frac{1}{2} + 0 \times \left(\frac{1}{2}\right)^2 + 1 \times \left(\frac{1}{2}\right)^3$$

would be written as 0.101. If we wanted to run through all the different possibilities, we'd see that there were two choices for the first
number, two for the second, two for the third, and so on *ad infinitum*.
In the end, this yields a grand total of

$$\overbrace{2 \times 2 \times 2 \times \ldots \times 2}^{\aleph_0 \text{ repetitions}} = 2^{\aleph_0}$$

different possibilities.

Cantor guessed that the continuum had to be the next aleph on the
list. In other words, he reckoned that $2^{\aleph_0} = \aleph_1$. The statement is known
as the *continuum hypothesis*, which you may remember as the first of
Hilbert's twenty-three unsolved mathematical problems from 1900. It
basically says that the continuum is one aleph up from the natural
numbers, although it isn't immediately obvious that this has to be true.
The continuum could just as well be the size of a higher aleph, or perhaps it has nothing to do with the alephs at all. Cantor became obsessed
with his hypothesis. His letters to Mittag-Leffler tell the story of an
increasingly desperate man. One moment he would write to Mittag-
Leffler in triumph, announcing that he had proven the hypothesis, and
the next he would write in despair, having identified a fatal mistake in
his work. He bounced between proof and disproof, between the illusion of success and the reality of failure.

To this day, the continuum hypothesis has never been proven, or
disproven. However, in 1963, the American mathematician Paul Cohen
made a significant discovery. Inspired by the work of Kurt Gödel, the
great Czech logician, Cohen showed that the continuum hypothesis
was independent of the basic building blocks of mathematics (the so-
called 'ZFC axioms', named, in part, after the mathematicians Ernst
Zermelo and Abraham Fraenkel). This means that the hypothesis
could be assumed to be either true or false, and neither would ever lead
to a contradiction. To understand this, imagine asking if a Liverpool
fan could also support their greatest rivals, Manchester United. You

would immediately realize that he couldn't since the two are in direct contradiction of one another. However, what if you were to ask if he also supported the Boston Red Sox? Given that the Red Sox play a different sport, there is no contradiction either way: it might be that he supports them, it might be that he doesn't. Cohen showed that mathematics was just as relaxed about the continuum hypothesis. Eighty years after Cantor had descended into insanity, Cohen was awarded a Fields medal for his work, the mathematical equivalent of a Nobel Prize.

In time, Cantor would see the continuum hypothesis as a matter of dogma, as transcending mathematics. The hypothesis belonged to God and, in Cantor's mind, God would protect it. In the latter half of his life, Cantor spent more and more time in the sanatorium. His breakdowns usually began explosively, ranting at the injustice of the world, until eventually, the depression would take hold. As his daughter, Else, would later recollect, he would become withdrawn and unable to interact. It was during his long recoveries that Cantor would often indulge another obsession, beyond his battle with the continuum. He would do battle with Shakespeare.

Cantor believed Shakespeare a fake, becoming convinced that his plays were the work of the seventeenth-century scholar Sir Francis Bacon. He was a native German speaker who also spoke Danish and Russian and, although English was his fourth language, he believed he knew it well enough to publish pamphlets in support of his radical Shakespearean hypothesis. In 1899, after another of his breakdowns, Cantor was granted medical leave from the University of Halle. This was then followed by a bizarre episode that offers a glimpse into Cantor's troubled state of mind. He sent a letter to the Ministry of Education, begging to be released of his professorship, to be left alone to work in a library in the service of the Kaiser. He told them he had extensive knowledge of history and literature, offering his pamphlets as evidence and even suggesting he had new information on the English monarchy and the identity of its first king. If the ministry did not respond to his request in a timely fashion, Cantor pledged to offer his services to the Russian Czar. As it happened, the ministry ignored his letter and Cantor never made contact with the Russians.

By the time the Great War was raging through Europe, the

importance of Cantor's mathematical work had been firmly established – in sharp contrast to his work on English literature. Conditions in Germany during the war meant his final days in the sanatorium were spent in poverty. When British polymath Bertrand Russell published his letters in 1951, he paid tribute to Cantor as 'one of the greatest intellects of the nineteenth century' whose 'lucid intervals were devoted to creating the theory of infinite numbers'. But he also added, 'After reading [his] letter, no one will be surprised to learn that he spent a large part of his life in a lunatic asylum.'

Cantor had dared to explore the heavens of infinity. His legacy was such that others would dare to look even further. It turns out there is a level of infinity that lies beyond even the infinitely infinite – these are numbers that are known as inaccessibles. To understand the idea behind inaccessibility, we first need to return to the finite realm – to the natural numbers. With the rules of arithmetic, is there ever a way to reach into the alephs? The answer is a resounding no. In the finite realm, all we ever have are finite numbers and all we can ever do is perform a finite number of sums, or multiplications or even exponentiations. As a result, our access to the alephs is denied. In this sense, $\aleph_0$ is an inaccessible cardinal number because we cannot reach it by playing arithmetic games on the finite cardinals below.

Now let's leap into the heavens.

Once we have our hands on $\aleph_0$, we know we can go out to even bigger cardinal numbers, through power sets and exponentiation. If the continuum hypothesis is right, $2^{\aleph_0}$ will immediately take us out to $\aleph_1$, and from there the rules of arithmetic allow us to go out to $\aleph_2$, $\aleph_3$, and so on. As we reach out to bigger and bigger cardinal numbers, we begin to wonder if there is anything we cannot access. The truth is we don't really know. One possibility is that the answer is no – that $\aleph_0$ is the only cardinal inaccessible to those below. But that is a rather dull conclusion. It is far more exciting to imagine higher alephs that are so large they are inaccessible to all the alephs that have gone before. So that's what mathematicians are inclined to do – after all, they get to make up their own rules and see what happens. With this perspective, let us consider the first of our new inaccessible numbers. From the realm of the lower alephs, we can look, but we cannot touch. No matter how often we exponentiate, we will never be able to reach it, just as we were unable to reach $\aleph_0$ from the finite realm. It is a new level of number, a celestial leviathan that eludes an infinity of infinities. It doesn't really have a name, so, in the spirit of Edward Kasner, whose nephew cooked up the googol, I asked my kids to think of one. In the end, they settled on *The Yeti*. I thought it was perfect. After all, the Yeti lives high in an inaccessible world and no one is sure if it really exists.

Do any of the alephs actually exist? Are they part of the physical realm? Cantor, limited only by his imagination, was able to walk with the infinite, to embrace it and understand it. But that was in the mathematical world of numbers and sets. In the physical world, the infinite is often considered a malady, a sickness that signals a lack of understanding, a calculational paralysis. But there are places where we have learned to overcome this paralysis, where infinity has been conquered and our physical theories have prospered. This is true of the infinities we encounter in electromagnetism and in nuclear physics. But it is not true of gravity. In gravity, there are an infinity of infinities. As we will see next, the paralysis is infinite.

CLOSE ENCOUNTERS
OF THE INFINITE KIND

Beware the tides of gravity. Beware the singularity at the centre of the black hole, where spacetime touches infinity. Beware the gravitational stress that grows and grows, that tears you apart, limb from limb, atom from atom, quark from quark. Beware your final moment, when time itself ceases to exist, and everything you are is consumed by the microscopic fabric of our universe.

This was the horror of Pōwehi, the black hole leviathan we encountered at the beginning of this book. We have seen it from afar, but what of its horror within? Is the singularity real? Is it really possible to touch *infinity*, albeit for a final moment in time? In 1965, the English mathematician Roger Penrose discovered something remarkable. If Einstein was right about gravity, then *every* black hole was an end, a cloak for a singularity, a facade for infinity. He showed that the singularity would always be there whenever there was a surface like an event horizon, beyond which no one could escape. Fifty-five years later, the elderly Penrose, by then a knight of the British realm, was awarded a Nobel Prize for this result. But despite the Swedish prize committee's stamp of approval, that doesn't mean that Penrose's singularities actually exist in nature. What Penrose's work actually showed was that if black holes exist, as we now believe they do, then Einstein's theory is broken. By harbouring this infinity, the theory is hiding something it simply cannot handle. In physics, infinity is a malady that has to be cured.

We've seen this sort of thing before.

There was a time, not that long ago, when our infinite maladies were far less exotic. They weren't just lurking inside black holes but in the glow of a light bulb or the buzz of a radio transmission. These mundane phenomena are scenes in the ballet of quantum electrodynamics, as photons dance with electrons and electrons dance with photons. The interaction between a photon and an electron is the most basic interaction in all of physics, and yet, in the build-up to the Second World War, it also appeared to be broken. The electronic dance was afflicted with an infinite malady.

The story began with Paul Dirac, my own academic grandfather, who connects to me through a direct line of PhD advisers. He was the son of a Swiss immigrant who had moved to Bristol, in the west of England, to teach French. He was a quiet boy and an even quieter adult. Colleagues in Cambridge introduced a unit of speech known as the 'Dirac' which counted as one word per hour. Dirac himself saw little need for words. He'd scoffed at Robert Oppenheimer's interest in poetry and claimed that his school had taught him not to begin a sentence unless he knew how it would finish. It was a school he'd shared with another boy who was far more loquacious – the Hollywood actor Cary Grant.

In 1927, Dirac proposed a theory that married Bohr's old ideas about quantized orbits of electrons in an atom with Einstein's ideas about relativity. It was the first ever quantum field theory and a major breakthrough in understanding the hustle and bustle of the microscopic world. He showed exactly how the electrons in an atom could communicate with the photons it released as radiation. Both the electron and the photon could be understood as quantum wiggles of *fields* – the electron a wiggle in the electron field, the photon a wiggle in the electromagnetic field. Each wiggle would trigger other wiggles, which triggered other wiggles, and so on. It was a work of such beauty that Dirac was afraid to explore its consequences, fearing that nature might have been foolish enough to choose something far less elegant.

There was initially great success. Powerful minds set about developing the idea into a new branch of physics known as quantum electrodynamics, or QED for short. They included a quartet of future Nobel Laureates: Pauli, Heisenberg, Fermi and the Hungarian Eugene Wigner, whose sister, Manci, would go on to marry Dirac. Together with Dirac, they began to unearth new and interesting phenomena from the creation and annihilation of particles in a magnetic field to the existence of antiparticles.

The early success of QED prompted speculation that they would soon be able to make predictions for all physical phenomena involving electromagnetic radiation and charged particles. However, the initial triumphs had come about by applying a technique known as *perturbation theory*. This is one of the most important tools in a physicist's locker. To understand how it works, let's put QED aside for a moment, and consider a more familiar scenario – the gravitational field of the Earth. To make our equations easier to solve, we normally treat the Earth as a perfect sphere. But the Earth is not a perfect sphere. Thanks to its rotation, it bulges out near the equator, giving a 1 per cent correction to the overall shape. Computing the effect of this on the gravitational pull is difficult to do exactly, so we approximate. We figure out the change in gravity to within the same 1 per cent accuracy using a bit of calculus and some funky theorems of mathematics. If we want to do even better, we work a bit harder and figure out the effect on the gravitational field at the next order in perturbation theory, working to an accuracy of 1 per cent squared or, in other words, one part in ten thousand. We could go on to an accuracy of 1 per cent cubed, or even higher powers. This is how perturbation theory works: you identify something small (in this case the 1 per cent change in the shape of the Earth), and you expand your results order by order in powers of that small parameter.

In QED there is also something small. This is the so-called fine structure constant, although most of us just call it alpha. It has nothing to do with the aleph and the omegas of the previous section. It is just a number that measures the strength of the interaction between photons and electrons – it tells us how much they want to dance. The value of alpha controls everything we see and much that we don't. It sets the size of atoms, the strength of magnets, the colours of nature. Its value has been measured to be very close to the fraction $1/137$, a fact which many physicists, past and present, have sought to better understand. Perhaps it was Pauli who was most obsessed. 'When I die,' he quipped, 'my first question to the Devil will be: what is the meaning of the fine structure constant?' Pauli would often dream of numerical coincidences, relating alpha to pi or some other important number. He even sought out the psychoanalyst Carl Jung, who analysed his dreams and became convinced that Pauli was gaining insight into 'some grand cosmic order'. In a bizarre coincidence, Pauli would eventually die of pancreatic cancer in Room 137 of the Red Cross Hospital in Zurich.

The smallness of alpha allowed those early dynamos of QED to calculate using perturbation theory. They began to work out the probabilities for different processes to occur, charged particles scattering this way and that, bouncing around photons, pushing and pulling in different directions. Their results were accurate to order alpha or, in other words, to within one part in 137, which is less than 1 per cent. To get even more accurate results, to within less than 1 per cent of 1 per cent, they just had to go to the next order in perturbation theory, to order alpha squared or even higher. It was only a matter of mathematical pain – nothing should actually go wrong.

It went wrong.

It started with Pauli. He realized that a lonely electron was not so lonely – it triggered an electromagnetic field. Whenever we bring a distribution of charge together into a small region of space, the electromagnetic field means we have to do work against the forces of repulsion. This means we have to feed energy into the system, and the smaller the region, the more work we have to do. This extra energy is known as the 'self-energy' and, in the case of the electron, you could think of it as contributing to the electron's mass (remember: energy and mass are equivalent). What upset Pauli was the idea that the electron was a point-like particle cramming all its charge into an infinitesimally small region. That would push the electron's self-energy, and therefore its mass, to infinitely large values.

Of course, Pauli understood that this wasn't quite right. One had to take into account quantum effects and, with the development of QED, he knew he had just the right theory to figure out what was really going on. He handed the task to his new assistant, a tall and fast-talking American who smoked far too many cigarettes. His name was Robert Oppenheimer.

Oppenheimer would go on to be the wartime head of the atomic weapons facility at Los Alamos Laboratory in New Mexico. Under his leadership, the team at Los Alamos successfully detonated the first atomic bomb in the desert of New Mexico on 16 July 1945. Less than a month later, the United States Airforce dropped two of these bombs on the Japanese cities of Hiroshima and Nagasaki, killing over two hundred thousand people. Quoting Hindu scripture, Oppenheimer later remarked, 'Now I am become Death, the destroyer of worlds.'

As a young physicist working under Pauli before the war, Oppenheimer had been known for his brilliance. But he was also known to be sloppy. 'Oppenheimer's physics is always interesting,' observed Pauli, 'but Oppenheimer's calculations are always wrong.' When Pauli asked him to look at the self-energy of the electron, Oppenheimer decided to study the problem in a concrete setting: he began to calculate the spectrum of light emitted by hydrogen atoms using QED. As ever, he had to resort to perturbation theory. At first it was a relatively straightforward problem. Working to order alpha, all he had to worry about was the proton in the nucleus exchanging a virtual photon with the orbiting electron. However, when he tried to figure out the corrections at order alpha squared, things started to get tricky. Oppenheimer realized there was a chance the electron and the photon could shapeshift. In particular, he had to worry about the effect of an electron emitting a photon before reabsorbing it a moment later. To his horror, Oppenheimer saw that the effect was infinite! It wasn't one of his calculational cock-ups – for once, he had done everything right. The problem arose because the transient photon could carry any amount of energy, rising all the way up to infinity. That meant he had to sum over all these possibilities. It was a sum he'd hoped would somehow conspire to give a finite answer, but that hadn't happened. QED had an infinite malady. With the distraction of a world war, it wouldn't be cured for almost two more decades.

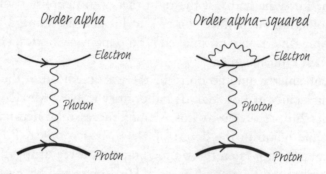

The proton in a hydrogen atom interacting with its electron. The left-hand image shows the physical effects to order alpha, corresponding to the exchange of a virtual photon. The right-hand image shows a correction at order alpha squared, with the electron emitting and reabsorbing another virtual photon.

Although the details were different, the problem was again one of an infinite self-energy, the electron picking up an infinite mass because of the way it interacts with its own electromagnetic field. Pauli was despondent. He said that he was tempted to leave physics and escape to the country to write 'utopian novels'. His melancholy obviously had a lasting impression on Oppenheimer. Rather than recognizing this infinity as a malady he could potentially cure, Oppenheimer saw it as a signal that the physics was wildly off course. Had he been more open-minded, he was as well placed as anyone to understand how the infinity could be tamed. Instead, that honour went to Schwinger, Feynman and the Japanese physicist Sin-Itiro Tomonaga.

To understand how these men finally conquered infinity, we return to Pauli's not-so-lonely electron. In addition to its electromagnetic field, the electron is surrounded by a sea of particles popping in and out of the vacuum – electrons, positrons and photons together in a

seething, bubbling virtual soup. There is no doubt about it: this soup
is going to affect the properties of the electron, including its mass. To
see why, imagine you hold a ping-pong ball under water and let go.
How much acceleration does it feel? The ping-pong ball is about
twelve times lighter than the water it displaces, which means the
buoyancy force is twelve times larger than the weight of the ball. If
this were all that mattered, the ball would experience an upwards
acceleration of 12g and the usual downwards acceleration of 1g, giving a net acceleration of 11g towards the surface. While the acceleration
is indeed upwards, you notice that it's nothing like that big. We need
to remember that the ball has to push some water out of the way. The
forces don't just accelerate the ball, they also have to accelerate the
surrounding fluid, and that makes it seem as if the ball is harder to
move. In the end, the ball acts as if it has more inertia or, in other
words, more mass. Physicists say that the mass of the ball is effectively
reconfigured, or 'renormalized', to a much larger value, so much so
that the upward acceleration ends up less than 2g. This renormaliza-
tion of the mass is a consequence of the fluid pushing back on the
ping-pong ball, interacting with it. It's the same with the virtual soup
that surrounds the electron. It pushes back on it, interacting with it,
'renormalizing' its mass. The difference between the electron and the
ping-pong ball is that the ball can eventually escape the water but the
electron will never escape the soup.

Oppenheimer used perturbation theory to carry out his calculation.
What this meant was that, to first approximation, it was as if there were
no quantum soup and the electron had the mass it would have in a 'soupless' classical world. When he calculated the first correction, it was as if
he were adding the soup. To his horror, he found that this correction was
infinite. In other words, the electron's new and improved 'soupy' mass
differed from the bare, soupless mass by an infinite amount. In the physical world, the electron is not infinitely heavy, so something seemed to
have gone disastrously wrong.

Except it hadn't.

What Oppenheimer failed to realize was that his calculation included
two different masses – the soupy mass and the soupless one – but only
one of these is physically relevant. The truth is you can only ever measure the soupy mass because the electron is *never* able to escape the

quantum soup. Oppenheimer had assumed that both masses needed to be finite for the theory to make sense, but that isn't the case: only the physical, soupy mass needs to be finite. The unphysical, soupless mass can never be measured, so it is fine for it to be infinite. In fact, it turns out it *has* to be infinite, at least as infinite as Oppenheimer's infinite quantum correction, but with the *opposite* sign.

Let's look again at the calculation soupless mass + quantum correction = soupy mass. If you like, if there is an 'infinity' in Oppenheimer's quantum correction, there must be a 'minus infinity' in the value of the soupless mass, in order to get a finite answer at the end. By themselves, these infinities are physically meaningless, so we don't get too upset by them. Of course, we never really use infinite values in any of our calculations, because we cannot keep those under control. Instead, we work with arbitrarily large but finite placeholders so the mathematics still makes sense. These placeholders – these proxies for infinity – are then assumed to cancel each other out. We are left with a finite value for the physical, soupy mass that matches the experimental measurements.

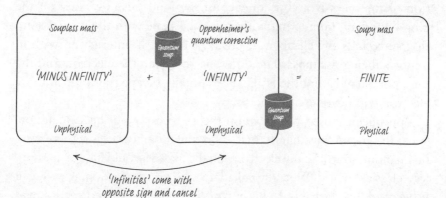

We could probably use an analogy. Let's imagine you are setting up a business buying and selling lollipops. The lollipops cost you £1 each, but you know you can sell them for double that on the first day's trade, although after that you will have to sell them at cost price. To get your business up and running, you borrow an infinite amount of money from a friend and buy an infinite number of lollipops. On the first day of trade, you sell a hundred lollipops. In monetary terms, how much are you actually worth? If we only looked at the bare value of your assets, we might assume you were infinitely rich. After all, you still have an infinite number of lollipops which you could sell at cost, as well as £200 in cash from the first day's sales. But this is only half the story. You still owe your friend the money you borrowed. When we subtract this debt, it's clear that you should just be left with the *profit* you made on the first day: £100. That's your true value.

In the analogy, the infinite value of your assets is like the infinite value of the electron mass in a soupless, classical world; the infinite debt is like Oppenheimer's infinite quantum correction; and the true level of your wealth (in this case, £100) is like the true physical value of the electron mass, surrounded by the quantum soup.

Before we declare the theory of the electron cured, we have to scan for other infinities. In QED, it turns out that the electron charge also picks up an infinitely large quantum correction. No matter. Just as we did before, we simply declare the bare, soupless charge to be infinite but impossible to measure. The infinite quantum correction once again comes with the opposite sign, the two infinities cancel, and we arrive at a finite soupy charge consistent with your experiments.

If that seems like a sleight of hand, now let's see the real magic.

You can use perturbation theory to calculate any process you like, electrons and photons bouncing around at random, and it all remains finite, as long as you stick to your guns and *declare* the soupy mass and the soupy charge to be finite. It seems like a miracle. Quantum corrections to a complicated process might include a plethora of infinite sums but, in the end, it doesn't matter. Those infinities are really just the remnants of the infinities we saw for the electron mass and the electron charge. Once the soupy mass and the soupy charge are fixed by experiment, *everything else falls into place*. There are no more infinities to worry about.

The infinite malady is cured.

In January 1948, Schwinger, who had not yet turned thirty, went through these ideas in front of a packed house at a meeting of the American Physical Society in New York City. Despite his tender years, he was already famous. He had enrolled at college at just fifteen years of age and, by the age of nineteen, he had seven research publications and was holding court with intellectual giants such as Pauli and Fermi. Ten years later, at the meeting in New York, he captivated his audience. His work was technically difficult, certainly, but it all worked out beautifully. The moment he fixed the soupy mass and the soupy charge to be finite and set by experimental measurements, he was able to calculate the effects on other processes and show that they too fitted the corresponding data. This included the way in which energy levels in hydrogen atoms were split by quantum effects – something that had been measured by Willis Lamb a year earlier, in 1947. From afar, it might have looked like he was playing far too fast and loose with infinity, but it didn't matter – Schwinger's masterclass was giving him all the right answers.

Feynman embarrassed himself that day. He had been working on similar ideas and, as Schwinger's presentation drew to a close, he informed the audience that he had recovered the same results. No one listened. Three months later, at a meeting in the Pocono Mountains in Pennsylvania, Feynman went again. He had a new and more intuitive way of thinking about QED. Everything was translated into pictures, cartoons where electrons were denoted with straight lines and photons with wiggly lines. These are the same cartoons we used to describe Oppenheimer's calculation of the spectrum of hydrogen. We didn't show it, but there is also a mathematical code for every line and every vertex that would allow us to carry out the same complex calculation in double-quick time. However, in 1948, only Feynman knew the code – no one else had any idea what his pictures really meant. Schwinger's methods were long and laborious, but at least they understood his language. Feynman claimed to have the same results, but nobody was really sure if that was true.

It was tough on Feynman, but it might have been tougher on Sin-Itiro Tomonaga. He had developed his ideas in 1943, a lone wolf working away in isolation in Japan, as the world was still at war. Four years later, Lamb had carried out his measurements of the energy levels of hydrogen, but Tomonaga learned of this only through an article in a Japanese newspaper. Realizing his theory could also reproduce the same data, he wrote to Oppenheimer, who quickly invited him to Princeton.

Three very different men appeared to be doing three very different things, but all of them were getting the same answers. It was an Englishman, Freeman Dyson, who sewed it all together. After taking a road trip with Feynman and patiently attending the lectures of Schwinger, he understood that everyone was on the same page – they were doing exactly the same thing, but in different ways. His epiphany came on a bus travelling through Nebraska. 'It came bursting into my consciousness, like an explosion,' he recalled. 'I had no pencil and paper, but everything was so clear I did not need to write it down.' In the end, it would be Feynman's methods that prevailed, once everyone got used to his diagrams. The infinite malady was cured or, as Feynman would say as he accepted his Nobel Prize in 1965, the infinities had been swept 'under the rug'.

Schwinger, Feynman and Tomonaga never actually walked in the infinite heavens, as Cantor had. As we hinted at earlier, their infinite gymnastics was only ever performed in the finite realm. If they had an infinite sum, they wouldn't consider the entire sum but some truncated version of it, something they could keep under control. For example, if they had to sum over an infinite range of energy, they might stop the sum at an arbitrarily large but finite value. If another infinite sum were to appear elsewhere in a different context, they could truncate it in the same way and happily compare the two. The hope was that these comparisons might even make sense as they restored the infinite limit. The truth is the three men weren't treating infinity as a number in the divine ways that Cantor had proposed but as a controllable *limit*. They weren't reaching for an infinite heaven – they were skirting around an infinite hell.

This pragmatic approach can also be extended to electroweak theory and the physics of the strong force. The infinite maladies are more difficult there, but they can still be cured in more or less the same way. None of these cures ever require us to think of infinity as anything more than a limit, and there is a good reason for this: the theories themselves are incomplete. For example, we know that QED can accurately describe the dance of a photon and an electron when the ballroom is the size of an atom, but what if it's a googol times smaller? Does QED still apply? Absolutely not. As we squeeze the size of the ballroom, we expect QED to give way to electroweak theory, and then to something else, as the particle dance moves to smaller and smaller distances and higher and higher energies. We now know that the infinities of QED arose because we imagined the theory would *always* hold, but it doesn't. No one is a hundred per cent sure what replaces QED down to infinitesimally small distances, but it doesn't actually matter. Schwinger and friends found a way to take controllable limits, navigating their way past the infinitesimals and infinity without having to know the details of what was really going on.

With these particular infinities now understood as limits, we're faced with a question: what about Cantor? Does his mathematics apply to nature, or is it supernatural? If Cantor's spirit is to be found anywhere in nature, it will surely be in the physics of quantum gravity. After all, in Einstein's classical model, gravity is a theory of the spacetime *continuum* – the same mathematical continuum that taunted Cantor for most of his life. What happens to it as we steer too close to a singularity, when quantum effects start to kick in? Does it become something else entirely, something that Cantor might also have seen in the infinite heavens?

We could try to build a quantum version of Einstein's theory from the bottom up, using perturbation theory, but we would soon run into serious trouble. Not only are there infinities, as there were for the other forces, there are infinitely many of them! That's a problem you cannot overcome. In quantum electrodynamics there were just two infinities to worry about – the electron charge and the electron mass. Once these were reconfigured to the finite values that were measured in experiments, everything else fell into place. When you try to quantize gravity in a similar way, to bring everything under control you soon realize you have to reconfigure an infinite number of different quantities. That requires an infinite number of inputs from an infinite number of measurements. By anybody's standards, that is not a working theory.

To really quantize gravity, you have to do something more radical. In loop quantum gravity, spacetime is pulverized, broken up into an uncountable number of building blocks – the so-called spin networks. The trouble is that it's not so easy to piece it all back together and, if you can't do that, you can't make contact with the basic empirical theory of gravity laid down four hundred years ago by Sir Isaac Newton. This is why most physicists, myself included, lean towards an alternative idea but one that is no less radical. It is not the rattle of a particle that echoes through the universe but the symphony of a string.

THE THEORY OF EVERYTHING

String theory is more than just a theory of quantum gravity. It is the theory of everything, the score for a universal waltz, directing the dance of electrons, photons, gluons, neutrinos, gravitons and all that exists in the physical world. And if our expectations are right, string theory is also a *finite* theory, the ultimate cure for the infinite maladies. Infinities are no longer swept under the rug, as in quantum electrodynamics. They are vanquished. They are completely absent. Cantor may have walked in the infinite heavens, but a string theorist simply doesn't have to.

It all began with the right wrong answer.

In the summer of 1968 the world was in chaos: war was raging in Vietnam, students were rioting in Paris and the assassinations of Martin Luther King and Bobby Kennedy had just convulsed the United States. At CERN, Gabriele Veneziano, a young Florentine physicist with a paradoxically Venetian name, was focused on the chaos of the microscopic world. He wanted to figure out what was happening when you took two hadrons and smashed them together.

We now know that hadrons, like the proton or the neutron, are made up of quarks, held together by the unbreakable bonds of the gluon. Although Murray Gell-Mann had already floated his idea of quarks in the early 1960s, at the end of the decade no one was entirely sure if they were real and the physics of hadrons was still not understood. In particle physics, whenever you bounce one particle off another and see what happens, you study a quantity known as the amplitude. This is just a complex number whose size tells you something about the probability for a particular process to occur. Veneziano was interested in two pions smashed together to produce a single pion alongside another hadron known as the omega (which obviously has nothing to do with Cantor's omega). His aim was to *guess* a mathematical formula for the corresponding amplitude – something that could reproduce the experimental data of the time and was mathematically consistent with both quantum mechanics and relativity.

To do this, Veneziano knew that he needed a mathematical function with some bespoke properties, but what was the function? Simple polynomials or trigonometric functions were not enough – he needed something a bit more sophisticated. Eventually, he found exactly what he was after hiding in the work of the great Swiss mathematician Leonhard Euler, who had lived two centuries before. After submitting his paper, Veneziano took a holiday in Italy and returned four weeks later to discover a flurry of excitement at what he'd achieved. It wasn't long before similar formulae were proposed for other hadronic processes. At some level, it was a game of mathematics, but when three of the world's most creative physicists – Yoichiro Nambu, Holger Bech Nielsen and Lenny Susskind – began to look at the equations a little more closely, they saw that something was wriggling.

Strings. Tiny, but eternally restless.

The three men covered a spectrum of personalities: Nambu, the shy Japanese; Nielsen, the unorthodox Dane; and Susskind, the charismatic New Yorker. But they also shared the creative spark that allowed them to see what was really going on inside Veneziano's formula. Each of them independently realized that Veneziano's amplitude could arise from a picture of hadrons as tiny rubber bands, as opposed to point-like particles. Those rubber bands are what we now think of as fundamental strings, extended in one direction, vibrating and wriggling in an infinite number of different ways. Veneziano had never imagined it this way, but he had inadvertently stumbled upon string theory. He had stumbled on the right wrong answer.

The strings are small, so small that they normally look like particles. It's only when you zoom right in that you notice they are extended a little. They can be open or closed, stretching between two different points in space or curled up in a loop. And when you pluck a string, it vibrates. This is where the music starts. Just as the different vibrations of a guitar string can yield different musical notes, so the vibrations of a fundamental string can mimic the effects of different particles. For example, the more frantic the vibrations, the more energy you have stored in the string. Because mass and energy are equivalent, it must be that the most frantically vibrating strings correspond to the heaviest particles.

In the early days of string theory, the spectrum of strings and the particles they were meant to represent started to become a concern. The worry was with the lightest strings. The lightest of all is the one that isn't plucked. You might think that it has vanishing mass, but that isn't true. In the last chapter we learned all about zero-point energies – the energies you get from unavoidable quantum wiggles. For a string, these turn out to be negative. When you work out the implications for the lightest string, you realize that it doesn't have negative energy – it has negative energy *squared*. That means the corresponding particle has an imaginary mass, proportional to the square root of minus one. This is known as a *tachyon* – a red flag for an instability. Tickling a tachyon into existence is a bit like nudging a pencil that is standing on its end. Everything just falls over. As far as the fledgling string theory was concerned, the tachyon had to be banished.

One level up from the tachyon, string theory was running into problems with experimental data. It turned out that strings that were only gently plucked had to have vanishing mass if they were to be compatible with relativity. They also had to have spin. That was a problem because string theory was designed to be a model for hadrons and, as experiments showed, there were no hadrons with these properties. Things got even worse when Claud Lovelace, a British-born physicist who had taught himself General Relativity and quantum mechanics aged just fifteen, made an alarming discovery.

In string theory, none of the dimensions of space and time are assumed to exist from the beginning – they actually *emerge* from the underlying theory. You begin with a fundamental string, stretched across just one dimension of space, and you imagine it filled with a number of fields that take different values at every point along the string. These fields can then encode the coordinates of the string in the full spacetime, and so the more fields you have, the more dimensions of spacetime there are in total. Lovelace realized that strings would be compatible with quantum mechanics only if there were twenty-six of these fields. In other words, spacetime had to have twenty-six dimensions. That's one dimension of time and *twenty-five* dimensions of space, a little more than the three-dimensional world you are almost certainly used to. As Lovelace would later remark, 'One has to be brave to suggest that spacetime has twenty-six dimensions.'

This was 1971 and, at around the same time, strings became superstrings. This was more than a lazy marketing strategy. String theory was enhanced with a fancy new symmetry – supersymmetry. We first came across this sort of symmetry in chapter '0.0000000000000001', when trying to control the mass of the Higgs boson. The details there are different, but the principle is the same: every fermion partners up with a boson and every boson with a fermion. For string theory, these partnerships brought about some improvements: the spacetime dimension was reduced from twenty-six to a measly ten dimensions and the tachyon was successfully banished. But it clearly wasn't enough. Strings were starting to lose their appeal. As a model for hadrons, they were being usurped by quantum chromodynamics. The data was beginning to show that protons, neutrinos, pions and all the rest of them were built from quarks and gluons in a kaleidoscope of colour. In the end, Veneziano's amplitude wasn't giving the right answers for colliding hadrons at higher and higher energies. String theory was pretty, but it was also pretty useless.

Or was it?

John Schwarz was a young American physicist who had been captivated by its beauty. At Caltech, he stumbled across a kindred spirit – the brilliant young Frenchman Joël Scherk. The two prodigies took another look at the lightest strings. The tachyon was gone, banished by supersymmetry, but what should become of the massless strings? Scherk and Schwarz had a remarkable epiphany; so too the Japanese physicist Tameki Yoneya, on the other side of the Pacific Ocean. The three men noticed that the massless strings looked an awful lot like the gluons of particle physics and the gravitons of General Relativity. Never mind hadrons, perhaps string theory was a theory of quantum gravity. Why, it might even be a theory of everything.

You'd think at this point that the world would have stopped, that every physicist would have rushed to strings, like prospectors panning for gold, eager to find some buried treasure. But this isn't what happened. In fact, string theory remained on the fringes for another decade. In the 1970s and early 1980s, intellectual heavyweights were more concerned with particle physics, making rapid advances in both theory and experiment. String theory was a sideshow. Its reputation wasn't helped when further investigations revealed a possible conflict with quantum mechanics, even in ten dimensions. Tragically, Joël Scherk would never see string theory triumph. Towards the end of the 1970s, he suffered a breakdown. He could sometimes be found crawling around the streets of Paris and would send bizarre telegrams to famous physicists, like Feynman, who he knew from Caltech. Aged just thirty-three years old, Scherk committed suicide.

The first string revolution would come in 1984. Schwarz was at the heart of it once more, this time in collaboration with the British physicist Michael Green (who would later go on to teach quantum field theory to the author of this book!). Green and Schwarz took on the subtle conflict between strings and quantum mechanics and showed it to be fake. String theory was back as a quantum theory of gravity and, this time, the world of physics took note. To many, string theory was fast becoming 'the only game in town'.

It soon became clear that there wasn't just one consistent formulation of string theory but five. The aim was to pick the right version, tickle it in just the right way and bingo – you would find the theory describing everything there is in our universe. The theory of everything should explain the origin of electrons, protons, neutrons and all the rest of the known particles with exactly the right mass, pushed and pulled in exactly the right way by the four fundamental forces of nature. But in the early days of string theory, it never quite turned out like that. In the end, the equations were always a bit too hard to handle. People approximated things here and there and saw hints of a universe similar to ours, but it was never enough. The only game in town wasn't as much fun any more. String theory was grinding to a halt.

It needed another revolution.

The second string revolution began on pi day, 14 March 1995. Ed Witten was lined up to talk first thing in the morning at a string-theory conference in Southern California. As he addressed the audience, he was quietly spoken with a higher than average pitch, but his words carried a profound intellectual authority. He was ready to storm the Bastille. Witten showed that the five different versions of string theory were describing the same physics in five different languages. When the equations got too hard in one language, he showed how they often became easier in another. With this deep insight, string theory was released from its calculational prison.

But Witten, as ever, went further.

He proposed a new theory – a mother theory – whose five daughters were the five different string theories we already knew about. He claimed that the mother was best understood in eleven dimensions of space and time where the fundamental objects are no longer strings but higher-dimensional membranes. This is M theory – a mystical eleven-dimensional theory that unifies the five different versions of string theory. Witten always meant that M was for membrane. However, others will say it stands for mother or magic or even mystery. The truth is we still don't know what M theory really is, not yet anyway.

There is a higher-dimensional elephant in the room.

Superstrings make sense only in ten dimensions of space and time; M theory is best understood in eleven dimensions. Wait. What are we saying? Forget quantum gravity – take a look around you. There are not ten or eleven dimensions, there are four: three of space and one of time. If there are supposed to be six or seven extra dimensions, then where are they?

They are hiding down the back of the sofa. They are on the end of your nose. You can even find them inside the queen's cucumber sandwich. They are everywhere, from here to Andromeda to the Evil Eye galaxy. But they are tiny, curled up out of sight, the silent perpetual partner that lives alongside our macroscopic world.

A dimension is really just a new direction of travel. When we say there are three dimensions of space, we mean there are three independent directions of travel: forward and backwards; left and right; and up and down. The six additional dimensions of string theory are just six new directions of travel. Because they are wrapped up like tiny little circles, you can't travel very far in these new directions before you get back to where you started. That's why you don't notice them.

To understand this a little better, let's imagine you are an ant. Not just any ant but the monstrous bullet ant, a giant from the lowland forests of South America. Scampering across the forest floor, you notice a stick lying in the dirt. As a good experimentalist, you decide to crawl along the surface of the stick to figure out how many dimensions it has. Certainly, you notice that you can move forward and backwards along its length, but you don't see that you can also go around the axis. 'The surface of the stick has one dimension!' you declare in triumph. But you are wrong. You are simply too big, too monstrous, to notice the circular direction. A black ant from the gardens of England would have fared much better. Being so much smaller, it would have noticed both dimensions of the stick – along its length and around its axis. In string theory, the six extra dimensions of space are said to be wrapped up small, just like the circular dimension of the stick. Like the bullet ant, we are simply too big to see them. We haven't even seen them at the Large Hadron Collider, even though we are peering into a world a billion times smaller than an atom. If the extra dimensions are there, they are dwarfed by everything we have ever seen in nature.

Although they are hidden away, the extra dimensions equip string theory with enormous potential. It turns out there are googols upon googols of ways of wrapping them up. The extra dimensions can be shaped like doughnuts or more exotic geometrical objects known as 'Calabi-Yau' surfaces, whose twists and turns render them almost impossible to imagine. You can fill the dimensions with magnetic flux or tie them up with strings and membranes. How you do the wrapping affects the physics of the macroscopic dimensions that are still left over. Wrap six dimensions up on a doughnut of some particular size and you will find a four-dimensional world filled with certain particles, pushed and pulled by a very particular set of forces. Wrap the extra dimensions on something more exotic and the world could look very different. String theorists like to work with these fancy Calabi-Yau surfaces because they don't destroy all the underlying supersymmetry – they leave a little bit left over in our four-dimensional world. We have already seen how supersymmetry can be useful for understanding why the Higgs is unexpectedly light or for unifying some of the fundamental forces. However, when we wrap up the extra dimensions in string theory, it plays another important role: it helps to keep the mathematics under control. Without it, the set-up is unreliable and the predictions of the theory can't always be trusted. The modern view is that string theory presents us with a multiverse – a landscape of different possible universes lined up alongside different Calabi-Yau surfaces, giving different particles, forces, vacuum energies and even different dimensionalities. It would appear as if our particular universe is just one of many possibilities.

But what of the infinite maladies that motivated our stringy ascent?

In string theory, infinity is vanquished. It is expected to be a finite theory, immune from the infinite curse that has afflicted particle physics since the 1930s. Although there isn't a watertight proof of this claim, there is good reason to believe it to be true. In particle physics, infinities arise because particles can kiss – they can touch one another. Those kisses allow particle pairs to pop in and out of existence over infinitesimally small times and infinitesimally small distances. It's like a frantic form of popping candy, firing the physics into the realm of infinite energies and infinite momentum. With strings, none of this can happen because they don't know how to kiss. Strings are extended in space, not by much, but enough to stop them kissing at a single point in space and time in the way that particles do. When strings come together, everything gets smoothed out. The popping candy is not as frantic, and the infinities are conquered.

You should feel good about this. String theory is the vaccine that ended the infinite plague. Tell your friends, tell your family, tell the bloke down the pub who has been banging on about loop quantum gravity. There is an inevitability to string theory, a continuous chain of ideas that began at the turn of the twentieth century with the twin pillars of relativity and quantum mechanics. It led us to the right wrong answers. Veneziano and his contemporaries weren't interested in tiny rubber bands. They were interested in amplitudes, mathematical formulae that respected the rules of the game, that were consistent with the pillars of physics. They didn't go looking for strings, but they found them, wriggling and writhing in the right wrong answer. They also found quantum gravity.

It is this intimate connection to relativity and quantum mechanics that also gives string theory its vulnerability. But that is a good thing. People often criticize string theory for being beyond experiment – that it can never be shown to fail, not even in principle, but that is simply not true. The principles that underpin relativity and quantum mechanics are being tested in experiments at this very moment in time. If those pillars were to fall, then string theory would also fall.

With infinity vanquished, what can string theory tell us about the true fate of the astronaut, torn apart by the tides of gravity, on his fateful journey towards the singularity of a black hole? The truth is we still don't know. The calculations are still too tough to figure this out, at least for the kind of black holes you expect to see in nature. To go further, we probably need another revolution: an insight into M theory, something that allows us to play with strings in the most violent of settings. This revolution promises to be the most profound discovery in the history of mankind, and with good reason. With the continuum of space and time twisted into oblivion, the singularity inside a black hole is not so different to the singularity of an infinite Big Bang. If the next revolution teaches us what is really happening deep inside a black hole, it may also teach us how the universe came into existence. It could teach us about Genesis itself – the singularity of our own creation.

And so it is, as we wonder about the beginning of time, that our story draws to an end. We have ridden on the back of fantastic numbers – big, small and heavenly infinite – and travelled through the fabric of the physical world. We have stood and admired the particles and strings that dance together in the microscopic ballroom, we have wrestled with leviathans, we have been humiliated by little numbers, we have seen ourselves as holograms on the edge of space, and we have journeyed to the most faraway corners of an unexpected world.

But in all of this, what did we really see? We saw the symbiotic relationship of mathematics and physics, how each of them thrives in the presence of the other. The synergy between maths and physics has never been more relevant to understanding how the universe is built. Our knowledge is now so deep that to see further with experiment can be technologically daunting and eye-wateringly expensive. For example, a particle collider ten times more powerful than the Large Hadron Collider at CERN is estimated to cost more than $20 billion. But we can also use mathematics to push the frontiers of physics. Right now, some people are trying to *mathematically* prove that string theory is the unique theory of quantum gravity. If they are successful, we no longer need to test string theory directly with experiment – we just need to test the assumptions that went into the underlying mathematics.

With mathematics, the physicist can dance, and with physics, the mathematician can sing. When we encountered the leviathans, the biggest and most magnificent numbers in the universe, we didn't just marvel at their size and the beauty of the underlying mathematics. We tried to comprehend them in our physical world. They gave us the opportunity to look upon the world at its most extreme. It was there, at the edge of physics, that the maths began to sing. It sang the sweet melody of relativity and quantum mechanics. It sang about the terror of Pōwehi. It sang about the holographic truth. When the little numbers taunted us with the mysteries of an unexpected world, the physicists danced the dance of symmetry. Or at least they tried to. They still haven't figured out all the steps.

Think of the fantastic numbers and let them sing in the fantastic world of fundamental physics. Think of 1.000000000000000858 and imagine yourself running alongside Usain Bolt, slowing down time like a relativistic wizard. Think of a googol and a googolplex and imagine a googolplician universe filled with doppelgängers, other versions of you and me, of Donald Trump and Justin Bieber. Think of Graham's number and experience the cerebral crush of black hole head death. Think of TREE(3) and imagine yourself playing the Game of Trees, far into the future of our universe, only to be halted by the cosmic reset, a timely reminder of a holographic truth.

Think of zero. Think not of its sin but of its beauty and the magic of symmetry in nature. Think of 0.000000000000000001 and 10^{-120} and see the mysteries of our universe, a chance to understand the unexpected nature of the Higgs boson and the energy of the cosmic vacuum. Think of infinity and remember Cantor's encounters with heaven and hell. Marvel at the symphony of physics and how infinity was conquered by the vibrations of strings.

Think of almost any number you like – there will surely be something wonderful about it, something fantastic. If you still don't believe me after everything you have read in this book, let me leave you with a century-old tale of two great mathematicians: the legendary number theorist G. H. Hardy and his Indian protégé Srinivasa Ramanujan. They were an unlikely couple. Hardy was a Cambridge don while Ramanujan had received no formal training in mathematics, growing up in Madras under British colonial rule. But Ramanujan was also a genius, a man who understood infinity, a man for whom mathematics was simply instinctive. In 1913, while still working as a clerk in the Department of Accounts of the Port Trust Office in Madras, Ramanujan sent a package of papers to Hardy, along with a covering letter asking Hardy to publish his work – he was too poor to publish them himself. When Hardy looked at what Ramanujan had written, he immediately recognized his brilliance and began a correspondence. The following year Ramanujan was on his way to England to collaborate with Hardy. He would stay there for the next five years.

Towards the end of his time in England, Ramanujan fell seriously ill with tuberculosis and vitamin deficiencies. When Hardy went to visit him in the sanatorium, he complained that his taxi number – 1729 – was rather dull. Hardy was worried that it was a bad omen, but Ramanujan was unconcerned. 'No, Hardy,' he replied, 'it is a very interesting number; it is the smallest number expressible as the sum of two cubes in two different ways':

$$1,729 = 1^3 + 12^3 = 9^3 + 10^3$$

The story offers more than just an insight into Ramanujan's remarkable mind. Sprinkle it with some twenty-first-century physics and we also get a glimpse into the fundamental structure of the physical world.

It starts with Pythagoras and his right-angled triangles. If the sides have length a, b and c, then we all know that they satisfy an equation of the form:

$$a^2 + b^2 = c^2$$

It is easy to find integer solutions to this equation. For example, $a = 3$, $b = 4$ and $c = 5$; or $a = 5$, $b = 12$ and $c = 13$. But what happens if we increase the exponent, so that we have equations like $a^3 + b^3 = c^3$ or $a^4 + b^4 = c^4$ or even higher powers? Can we still find integer solutions? Around 1637, a French mathematician by the name of Pierre de Fermat confidently declared that the answer was no. In the margins of a copy of Diophantus's *Arithmetica*, he wrote:

'It is impossible to separate a cube into two cubes, or a fourth power into two fourth powers, or in general, any power higher than the second, into two like powers. I have discovered a truly marvellous proof of this, which this margin is too narrow to contain.'

The claim is, of course, true, but it famously wasn't proven until the mid-1990s, by the English mathematician Andrew Wiles. Almost eighty years earlier, Ramanujan had set about *dis*proving it and, in doing so, he came across the number of Hardy's taxi – 1729. Ramanujan's idea was to cook up a counterexample to Fermat's claim. We now know that this was impossible, which explains why he had to contend himself with a family of near misses. As you can see, $9^3 + 10^3$ gives $1,729$, which is *almost* the same as 12^3, missing the target by one. He also noticed that $11,161^3 + 11,468^3$ is just one more than $14,258^3$, and that $65,601^3 + 67,402^3$ is one more than $83,802^3$. In fact, he figured out a way to find an infinite number of similar examples where the target is missed by a single unit.

Well, this story doesn't end with a failed attack on Fermat's last theorem. As it happened, Ramanujan's method gave him solutions to certain equations containing cubic powers and rational numbers. The scholar Ken Ono was one of those who uncovered the work in Ramanujan's famous lost notebook, which had been hidden for over half a century in the Wren Library at Trinity College, Cambridge. When Ono and his Ph.D. student Sarah Trebat-Leder began looking at the equations more closely, they realized that Ramanujan had been dabbling in a special family of geometric structures known as K3 surfaces. Interest in these strange and wonderful higher-dimensional shapes had blossomed long after Ramanujan's death, in the late 1950s, while his work still lay undiscovered. The name K3 was in honour of three other mathematicians – Kummer, Kähler and Kodaira – who had worked on closely related topics, and the deadly K2 mountain in the Himalayas. The climber George Bell had once described K2 as a 'savage mountain that tries to kill you'. K3 surfaces can be just as hostile, at least in the eyes of those mathematicians brave enough to study them.

But what does any of this have to do with the physical world?

It turns out there is a very good reason to take on this savage branch of mathematics: K3s are the prototype for the Calabi-Yau surfaces we mentioned earlier – the tiny, exotic shapes most string theorists use to squirrel away our extra dimensions. These are the shapes that control the physics of our macroscopic world. Hardy had complained that the number $1,729$ was rather dull, but the truth is he couldn't have been more wrong. It is intimately connected to the extra dimensions that hide silently alongside each and every one of us, that determine why the universe is what it is and why we are who we are.

No, Hardy, $1,729$ isn't dull. It's bloody fantastic. Just like every other number.

Notes

1.00000000000000000858

1 Strictly speaking, 299,792,458m/s is the speed of light *in a vacuum*. In the presence of a medium such as air or glass, light can be slowed a little, although this has nothing to do with relativity. Light only *appears* to move more slowly in these dense environments because it is constantly absorbed and re-emitted by the atoms or molecules that make up the material.

2 If the relative speed is v, then it turns out that time is slowed by a factor of $\gamma = 1 / \sqrt{1 - v^2 / c^2}$, where c is the speed of light, taken to be 299,792,458m/s. If v *is close to* the speed of light, time is slowed by a huge amount, close to a standstill. For Usain Bolt travelling relative to the stadium in Berlin at 12.42m/s, time is slowed by a factor of 1.00000000000000000858.

3 Why the c^2 in $E = mc^2$? Energy and mass are known to differ by units of 'speed squared', so the extra factor of c^2 helps ensure that the units match on both sides of the equation. It's a bit like trading dollars for pounds. But why c^2 and not $3c^2$ or $0.5c^2$? Well, if Bolt starts moving, we expect to pick up some kinetic energy so that we now have $E = mc^2 + \frac{1}{2} mv^2$ but, as usual, this is just approximating a missing factor of $\gamma = 1 / \sqrt{1 - v^2 / c^2}$ from special relativity – the correct expression is $E = mc^2 / \sqrt{\left(1 - v^2 / c^2\right)}$. This only works if the leading order part is precisely mc^2.

4 Since $x / t = c$, we rearrange and plug $x = ct$ into Minkowski's formula for spacetime distance. This gives $d^2 = c^2 t^2 - c^2 t^2 = 0$.

5 This description is adapted from *Albert Einstein and His Inflatable Universe*, by Mike Goldsmith (Scholastic, 2001).

6 *The Young Centre of the Earth*, by U. I. Uggerhøj, R. E. Mikkelsen and J. Faye, in *European Journal of Physics* 37, 3, May 2016.

7 For a non-rotating black hole, the innermost stable circular orbit of any planet or star lies at a radius one and a half times the size of the radius of the event horizon. For rotating black holes there are stable circular orbits around the equator whose trajectories appear to edge closer and closer to the horizon as we ramp up the spin. There is a maximum possible spin set by the black hole mass and for such maximally rotating black holes the innermost stable orbits can almost graze the event horizon.

A GOOGOL

1 This recursive naming definition was proposed by the celebrated googologist Jonathan Bowers, also known as Hedrondude.

2 *Our Mathematical Universe: My Quest for the Ultimate Nature of Reality*, by Max Tegmark (Alfred A. Knopf, 2014).

3 A modern version of Clausius's formula states that $\Delta S = \dfrac{\Delta E}{kT}$ where ΔE is the change in energy, ΔS is the change in entropy, T is the temperature and k is the so-called Boltzmann constant. In everyday units, k is rather tiny, equivalent to 1.38×10^{-23} joules per kelvin. Clausius didn't include Boltzmann's constant in his original formula. For him it was secretly absorbed into his definition of entropy.

4 During his *annus mirabilis* of 1905, Einstein showed how random molecular collisions in Bernoulli's model could explain *Brownian motion*, the jagged, lifelike movement of tiny grains suspended in a fluid.

5 In the mid-1990s, Andy Strominger and Cumrun Vafa of Harvard University managed to identify the microstates for a highly specialized, and somewhat artificial, family of black holes in string theory. When they counted them up, they were able to recover Bekenstein and Hawking's formula for the entropy.

6 For this black hole, the horizon area, $A_H \sim 1$ metre, and since $l_p \sim 10^{-35}$ metres, the entropy comes to $1 / \left(4 \times \left(1 \text{ metre}\right)\right)^2 / \left(1.6 \times 10^{-35} \text{ metre}\right)^2 \sim 10^{69}$.

A GOOGOLPLEX

1   This analogy is taken from *The Elegant Universe* by Brian Greene (Vintage, 1999).

2   At a temperature, T, it turns out that each snake pair carries, on average, kT worth of energy. The oven is heated to 180 degrees Celsius, which is 453 kelvin, and given that $k = 1.38 \times 10^{-23}$ joules per kelvin, we get an average energy of $kT = 1.38 \times 10^{-23} \times 453$ joules $= 6.25 \times 10^{-21}$ joules. Or, in other words, about 6 zeptojoules.

3   Notable experiments measuring the energy radiated by hot objects were carried out by the German physicists Lummer, Kurlbaum, and Pringsheim at the end of the nineteenth century. Of course, Lummer and his collaborators weren't measuring the radiation in your oven but other sources of similar radiation, including an electrically heated platinum cylinder.

4   De Broglie argued that a particle of momentum p could be associated with a wave of wavelength $\lambda = 2\pi\hbar/p$. For a photon for which we know the angular frequency, ω and energy, E, this works as follows: for the basic chunks, we know we can relate $E = \hbar\omega$, but since the photon travels at light speed, we also have that the momentum $p = E/c$ and the wavelength $\lambda = 2\pi c/\omega$. Putting all three together, we get $\lambda = 2\pi\hbar/p$. De Broglie simply extended this wave formula to all particles.

GRAHAM'S NUMBER

1   This explanation is adapted from *Numericon* by Marianne Freiberger and Rachel Thomas (Quercus, 2015).

2   Mathematicians normally refer to Ramsey numbers in terms of a pair of integers n and m, where $R(m,n)$ is the smallest number of people you need to invite in order to get a clique of m friends *or* a clique of n strangers. However, to keep things simple, I will always refer to $R(n, n)$ as the nth Ramsey number.

3   A typical speck of house dust has a mass of around a microgram. To accumulate the same mass in data we would need to store $10^{-3}/10^{-26} = 10^{23}$ bits. There are 8 bits in a byte, so that comes to around 10^{22} bytes, or 10^{13} gigabytes.

4   My iPhone is made from 31 grams of aluminium, which represents about a quarter of its weight. To get a ballpark figure for the total entropy, we compute the entropy stored in aluminium. Aluminium has a standard molar entropy (measured in units of Boltzmann's constant) of 28.3 joules per mol per kelvin. In the dimensionless units of entropy we are using, that equates to 2×10^{24} nats per mol. Aluminium has a molar mass of 26.98

grams per mol, so 31 grams of it must carry $31 \times 2 \times 10^{24} / 26.98 = 2.3 \times 10^{24}$ nats. Extrapolating to the entire phone, we estimate a total entropy of 10^{25} nats, or around 10^{15} gigabytes.

5 The surface area of my iPhone is around 19,000 mm². A black hole whose horizon is the same size would, by Hawking's formula, have an entropy of around 2×10^{67} nats. That corresponds to roughly 10^{57} gigabytes.

6 Susskind's entropy limit isn't perfect. Although it applies to a spaceship or an egg, there are some extreme situations where it fails, like a collapsing star or a spherical universe. A more sophisticated entropy limit has been developed by Berkeley physicist Rafael Bousso, which seems to work in all situations, including these more exotic cases.

ZERO

1 Translations correspond to shifting every part of an image by a fixed amount in a fixed direction, just as we see in the repetitive rows of kernels of corn or the glittering staccato of scales on a fish. Glide reflections are more exotic. You can think of them as a shift followed by a flip. Because of how we walk, a line of human footprints is automatically mapped out by a series of glide reflections. To see this, take a walk along some wet sand. Notice how your left print is exactly the same as your right, as long as you shift it forward slightly and flip it over.

2 See Chapter 3 of *The Symmetries of Things* by John H. Conway, Heidi Burgiel and Chaim Goodman-Strauss (A. K. Peters/CRC Press, 2008).

3 See *Zero: The Biography of a Dangerous Idea* by Charles Seife (Viking Adult, 2000).

4 We can prove this as follows: let $x = 1.111\ldots$ with a recurring sequence of ones, and multiply through by ten to give $10x = 11.111\ldots$ where again there is a recurring sequence of ones. Now subtract $10x - x = 11.111\ldots - 1.111\ldots$. The recurring ones cancel, giving $9x = 10$, and so $x = \dfrac{10}{9} = 1 + \dfrac{1}{9}$.

5 With such uncertainty surrounding the date of the Bakhshali manuscript, perhaps we should look to Cambodia for the earliest zero. It can be seen as a dot on an Old Khmer inscription carved on to a stone tablet dated AD 683. The date is considered authentic. The Old Khmer region had strong cultural links to the Indian subcontinent so this particular zero retains the connection to Indian mathematics. The ancient tablet was originally discovered in the late nineteenth century but went missing for

many years, following the murderous reign of the Khmer Rouge. It wasn't rediscovered until 2013, when the author Amir Aczel found it gathering dust in a shed at Angkor Conservation.

0.0000000000000001

1 Particle physicists like to talk about energies in units of eV, or 'electron volts'. 1 eV is the kinetic energy gained by an electron after you've accelerated it through a potential of one volt. By Einstein's famous formula, $E = mc^2$, we can relate this energy to an equivalent mass of around 1.78×10^{-38} kg. Electron volts are good for measuring the tiny masses and energies of fundamental particles. They are less good for everyday objects like humans. No one likes to be told that they weigh around 40 trillion trillion trillion electron volts – eleven stone sounds much better.

2 A very canny reader might be wondering how this lines up with the doppelgänger search in chapters A Googol and A Googolplex. There we described your doppelgänger as being an exact replica in the exact same quantum state. Given the presence of fermions, that might seem in contradiction with Pauli's principle. However, your doppelgänger is very far away, so far that it cannot be thought of as the same quantum system, so there is no contradiction.

3 To understand the jargon associated with left- and right-handed particles, it is useful to think about turning a screw. My wife (who is much better than me at DIY) once told me about the following mnemonic: righty tighty, lefty loosey. In other words, if you turn a screw clockwise, it moves forward, and if you turn it anticlockwise, it moves backwards. With an electron, we say it's right-handed if the spin would take the screw forward, in the same direction as the particle, and it's left-handed if the spin would take the screw backwards, in the opposite direction.

4 Weinberg and Glashow shared the 1979 prize with Abdus Salam, a well-known Pakistani physicist who had independently developed similar ideas to Glashow when he was working with the Englishman John Ward. As there can never be more than three winners for a Nobel Prize in any one year, Ward missed out and responded to the snub by sending a telegram to Salam that read 'Warmly Admired, Richly Deserved'. Check out the first letters of each of the four words.

5 As well as the 2013 Nobel Prize awarded to Higgs and Englert, Kibble also made a significant contribution to the 1979 prize won by Glashow, Salam and Weinberg. Kibble's work in understanding spontaneous

symmetry breaking in more complex settings was an essential ingredient in the development of electroweak theory.

6 The huge extremes in energy can be worked out by pushing the uncertainty relation to the limit. For an interval as short as the Planck time $t_{pl} \approx 5 \times 10^{-44}$ seconds, we can reach energies as high as $E_{max} = \dfrac{\hbar}{2t_{pl}}$. This comes to around a billion joules, given that Planck's constant, $\hbar \approx 10^{-34}$ joule seconds. We can now use Einstein's formula, $E = mc^2$, to convert this into an equivalent mass of around 11 micrograms – the mass of a quantum black hole! This is actually a pretty good estimate for the amount of mass that is fed through to the Higgs. As it happens, a more sophisticated textbook calculation puts the wannabe Higgs mass at a slightly smaller value, $\dfrac{1}{\sqrt{2\pi^2}} \times 11$ micrograms $\oplus 2.5$ micrograms, closer to the mass of a fairyfly.

10^{-120}

1 See *Nullpunktsenergie und Anordnung nicht vertauschbarer Faktoren im Hamiltonoperator* by C. P. Enz and A. Thellung, *Helvetica Physica Acta* 33, 839 (1960).

2 To compute the energy in each box, we just max out the uncertainty principle, using $E_{max} = \dfrac{\hbar}{2t_{min}}$, for a shortest time $t_{min} \approx 10^{-23}$ seconds. Without our modern understanding, Pauli would have done the calculation slightly differently. Although it was never published, it's thought his estimate was based on an idiosyncratic model of quantum theory that had been proposed by Planck a decade or so earlier.

3 Heisenberg was able to describe quantum mechanics with very few working parts, albeit in a complex mathematical framework. In contrast, by introducing the wavefunction, Schrödinger added an extra ingredient that simplified the machinery but which has been prone to over-interpretation. The wavefunction is often imagined as physically real, as real as a classical electromagnetic field, but this isn't correct. It is simply a means to an end – a way to encode probabilities and the possible outcomes of experiments. It is not something that can be measured directly in experiment.

4 See *Likely Values of the Cosmological Constant* by Hugo Martel, Paul R. Shapiro and Steven Weinberg, *Astrophysical Journal* 492, 1.

Acknowledgements

64. That's another *Fantastic Number*. In fact, it's a *dodecagonal* number, which is a bit like a triangular or square number adapted to a dodecagon. It's also the number of people I'm about to thank for helping me get this book off the ground. Of course, 64 doesn't really reflect the number of people who actually helped me. It is certainly an underestimate, and one that falls seriously short, a bit like the estimate Friedmann offered up for TREE(3).

I'll begin with Hendo.

My mate.

A few years ago he told me he was seriously ill with cancer. I, like many others, didn't want to accept it. We were ready to get Hendo whatever he needed to get better, so we began to raise money. I started giving public talks about Fantastic Numbers to large groups up and down the country, asking for donations from the people who came to listen. I raised a few grand that way. Between all his friends and family, we raised around £200,000, but it wasn't enough. We couldn't save Hendo. He left us, and we miss him terribly.

But something good did come from those public talks. I realized they could be the seed of a book. This book. A book that would only be completed thanks to the support of all my friends and family. I'll start with the little ones: my two lovely daughters, Jess and Bella, always cheeky, always calling me 'Gilderoy' if I ever get too big for my boots. To be honest, my wife, Renata, encourages them. She also encourages me. She was the first to read every word I wrote, always giving me honest and insightful feedback. I don't know how she did it really, because she isn't really that interested in science – she prefers *Bake Off*. But somehow she made sure I didn't send anything on to

the publishers with a 'soggy bottom'. So thank you, Perks, for everything x.

Thanks also to my mum and dad, always there and always believing
in me, as well as my brother, Ramón, and my sister, Susie. Thanks to
my in-laws, Cathy, Graham, Bob, Wendy, Austin and Mike, my old
mate Neil and, of course, all my nephews and nieces: my god-daughter,
Kirsten, Adam, Air Commander Elliot, Liverpool's next star Lucas,
Lyla, Jude, Jago and Hattie. I hope you all read this one day because
I'll be setting an exam. Extra special thanks to Adam for helping me
bounce around philosophical ideas in chapter 'Zero'. I hope he becomes
a philosopher one day.

I have to give a huge thank-you to Will Francis, my agent, and
everyone else at Janklow and Nesbitt. They have been incredibly sup-
portive, from helping me cobble together a meaningful book proposal
to scouring the world for new deals and new opportunities. Will has
always got my back.

Thanks to my editors, Laura Stickney and Sarah Day from Penguin
and Eric Chinski from FSG. I worked really closely with Laura, and
her input has improved the manuscript beyond all recognition. I'm
new to this game, and I guess at times it must have showed. Laura's
experience helped me arrive at something I hope we can both be
proud of. The support from everyone at Penguin and FSG has been
magnificent throughout.

Thanks to everyone else who read bits of the manuscript and told
me which parts were good, and which needed to go: Smarty, Bellars,
Norrie, Deano and Burrell, Ian from across the road, my father-in-
law, Bob, my colleagues, Ed Copeland, Pete Millington and Florian
Niedermann and my student Robert Smith. Extra special thanks goes
to my other student, Cesc Cunillera, a brilliant young mathematical
physicist, who went through the entire book checking every fact and
re-doing every calculation. He says I got the answers right – most of
the time, anyway.

Thanks to all the other friends and family who got a mention in this
book. Sadly, one of them, my neighbour Gary, who I mentioned
towards the end of chapter 'Zero', is no longer with us. Without him,
our street isn't quite as fun as it used to be.

Thanks to Ruth Gregory and Nemanja Kaloper for shaping me as a mathematician and as a physicist. Thanks to all the people whose advice I sought in writing this book, whether it be questions about physics, maths or the quirks of ancient Greece: Omar Almaini, Tasos Avgoustidis, Steven Bamford, Clare Burrage, Andy Clarke, Christos Charmousis, Frank Close, Gia Dvali, Pedro Ferreira, Ingrid Gnerlich, Anne Green, Stephen Jones, Helge Kragh, Juan Maldacena, Phil Moriarty, Adam Moss, Lubos Motl, David Pesetsky, Paul Saffin, Thomas Sotiriou, Jonathan Tallant and James Wokes. Thanks also to all the other people who have written the amazing books and articles, text and trade, that I used as inspiration.

And then there is Brady Haran.

I'm under no illusion as to the role *Sixty Symbols* and *Numberphile* have had in giving me the opportunity to fulfil this particular dream. Making videos with Brady has always been a lot of fun. He loves to throw you a curve ball, just as you are in full flow describing the wonders of the mathematical universe. Brady has given me the platform to reach out with my mathematical ideas, and he's still teaching me how to do it properly.

I want to finish with another number. As a Liverpudlian, there is one number that matters more than any other.

It's the number 97.

May they rest in peace and may their families get the justice they deserve.

Index

A Note About the Author

Antonio Padilla is a leading theoretical physicist and cosmologist at the University of Nottingham. In 2016, he and his team shared the Buchalter Cosmology Prize for their work on the cosmological constant. He is also a star of the Numberphile YouTube network, where his most popular videos include a discussion of Ramanujan's sum of all positive integers, which has been viewed more than eight million times.